GEOGRAPHIC INFORMATION
MANAGEMENT IN
LOCAL GOVERNMENT

GEOGRAPHIC INFORMATION MANAGEMENT IN LOCAL GOVERNMENT

IAN GILFOYLE

AND

PETER THORPE

CRC Press
Taylor & Francis Group
Boca Raton London New York

CRC Press is an imprint of the
Taylor & Francis Group, an **informa** business

CRC Press
Taylor & Francis Group
6000 Broken Sound Parkway NW, Suite 300
Boca Raton, FL 33487-2742

First issued in paperback 2019

ISBN-13: 978-0-7484-0935-8 (hbk)
ISBN-13: 978-0-367-39419-6 (pbk)

This book contains information obtained from authentic and highly regarded sources. Reasonable efforts have been made to publish reliable data and information, but the author and publisher cannot assume responsibility for the validity of all materials or the consequences of their use. The authors and publishers have attempted to trace the copyright holders of all material reproduced in this publication and apologize to copyright holders if permission to publish in this form has not been obtained. If any copyright material has not been acknowledged please write and let us know so we may rectify in any future reprint.

Trademark Notice: Product or corporate names may be trademarks or registered trademarks, and are used only for identification and explanation without intent to infringe.

Library of Congress Cataloging-in-Publication Data

Gilfoyle, Ian.
　　Geographic information management in local government / Ian Gilfoyle, Peter Thorpe.
　　　　p.　cm.
　　Includes bibliographical references and index.
　　ISBN 0-7484-0935-1 (alk. paper)
　　1. Geographic information systems—Government policy—Great Britain. 2. Information storage and retrieval systems—Geography—Government policy—Great Britain. 3. Local governments—Great Britain. I. Thorpe, Peter, 1943-II. Title.

G70.215G7G55 2004
910'.285—dc22　　　　　　　　　　　　　　　　　　　　　2004043851

Library of Congress Card Number 2004043851

Visit the Taylor & Francis Web site at
http://www.taylorandfrancis.com

and the CRC Press Web site at
http://www.crcpress.com

Foreword

Local government was an early pioneer of digital information. As intensive users of mapping for a host of public services on a daily basis, many local authority officers recognized the potential of digital geographic data ahead of their central government colleagues and, indeed, ahead of many operators in the private sector.

Such was the momentum within local government that by 1993 Ordnance Survey, Great Britain's national mapping agency, had signed a national agreement with the local government community to supply data to any local authority, police authority, fire service, and national park that could make use of it. Within a few years, every principal council in the country was using digital mapping data in some form somewhere within their authority.

Being early adopters of GIS created an expert class of user, but inevitably it also led to a legacy of older systems, processes, and — because information may have been collected over long periods of time and to different criteria — out-of-date and often incompatible datasets. Frustratingly, original systems were often established for very specific purposes, making their corporate-wide extension difficult when the wider potential was recognized.

However, local government is now empowered with a more strategic policy decision-making role at a local and community level, and at the same time being required to meet a wide range of central government targets. Policies and initiatives such as e-government and 21st century government, comprehensive performance assessment, best value, beacon status, local public service agreements (PSAs), modernizing government, and joined-up government and information are all now firmly on the agenda at the highest level.

The way in which local government targets its resources and funds activities to achieve its public service delivery goals is more widely scrutinized, monitored, and measured than ever before. As a consequence, the true value of geographic information is now being appreciated at senior management level, with the use of GIS being widened from that of being a specific technical or operational tool for a particular service to one where greater corporate benefits are being sought and achieved — often through innovative applications.

The challenge now is to effectively unleash the potential of geographic information by making the best use of investments — to "sweat the assets." This is not just about the appropriate use of the latest technology and obtaining up-to-date data, but fundamentally getting the right data and information management strategies, cultural changes, and organizational structures and processes in place, all in conjunction with appropriate corporate sponsorship and staff buy-in.

Key decisions have to be made, for example, on whether to employ a corporate GIS platform or a data and systems architecture that supports multiple vendor GIS solutions, and also how to benefit from the introduction of new technologies such as Web services. Also, decisions have to be made whether data captured by local government and others are to acceptable standards of accuracy and currency in today's world of developing technologies such as GPS. In addition, consideration must be given as to how best to manage all the data held locally by local government so it can be successfully shared both within departments and across government. Finally, how do all these decisions help to create effective citizen based services?

This book is a timely contribution to this process in which continued education, best practice guidance, and effective partnerships are key challenges to the wider GIS industry.

Effectively, GIS need to evolve from mapping and visualization tools to tools that provide valuable management information — cross organization or cross-sector information sharing based on a common and unique geographic reference — for joined-up government, emergency planning, streetworks, community portals and so on, a kind of geographic DNA in a way. GIS must evolve as tools for trend and predictive analysis, e.g., for crime prevention, neighborhood renewal and regeneration, highways and integrated transport planning, environmental impact analysis, raising educational standards, incident response, child safety, etc. Ultimately, they need to be evidence-based strategic decision-making tools to drive policy, grant allocation, community strategy, and public service delivery.

GIS are evolving to become true multichannel information systems. They can provide local government with cost-effective mechanisms for other authorities, partners, central government, the private sector, and the public to view, access, consult, and share information. They can also become customer service (eCRM) tools that facilitate choices in the way the public gain information via different contact channels and devices, e.g., call center, Web, interactive television, mobile phone, kiosk, PDA, information center, or shop.

I commend the publication of this book. Its structure and practical approach to the subject, including its use of highly relevant case studies for different types and sizes of authority, should do much to reinforce and advance the message that geographical information is an immensely powerful tool that can bring enormous benefits to local authorities and the people they serve.

Vanessa Lawrence
Director General and CEO, Ordnance Survey Great Britain

Acknowledgments

It is a sobering thought for us to realize that the origins of this book lie in the last century. When we first met Tony Moore, a Taylor & Francis senior editor, in the late 1990s, we naively expected that from first putting pen to paper — or fingers on the keyboard — to final publication would take about 18 months, and certainly no more than 2 years. Tony cautioned that our expectations were too ambitious and that it would take much longer than that to research the subject material, write all the chapters, and then finalize the manuscript.

How right he was! Here we are some 4 years later, putting final touches to a book that has been both stimulating and rewarding to write and hopefully will be interesting and instructive to our readers. Along the way, many people have helped with support, advice, and ideas, and we would like to take this opportunity to thank them all. We had early encouragement from Cecilia Wong (now Professor Wong at Liverpool), Sarah Lindley and Bob Barr from Manchester University, Vanessa Lawrence (then with Autodesk), and Andy Coote (ESRI). We also owe a debt of gratitude to Andrew Larner and his colleagues from the IDeA who helped us write Chapter 7; to Mark Linehan, the AGI director; and especially to Professor Michael Batty of CASA who contributed most of the final chapter.

Many of the ideas in this book stem from the experiences of the various case studies. We gratefully acknowledge the help of the following persons who not only provided information for these case studies but also commented upon them while in draft:

Mike Somorjay and Chris Butler of Bristol City Council, Bruce Yeoman of Bruce Yeoman Associates, and Tim Musgrave of TerraQuest Information Management, in relation to the case study of Bristol City in Chapter 9

Nick Adnitt, formerly of Southampton City Council, in relation to the case study of Southampton City Council in Chapter 10

Peter Shilson of Leeds City Council, in relation to the case study of Leeds City Council in Chapter 11

Ron Hillaby and Bill Taylor of Newcastle City Council, in relation to the case study of Newcastle City Council in Chapter 12

Malcolm Baker, formerly of Aylesbury Vale District Council, in relation to the case study of Aylesbury Vale District Council in Chapter 13

Steve Dean of Shepway District Council, in relation to the case study of Shepway District Council in Chapter 14

Mick Wooden of London Borough of Enfield, in relation to the case study of London Borough of Enfield in Chapter 15

Ian Pearce and Stephen Forgan of London Borough of Harrow, and Felicity Holland, formerly of London Borough of Harrow, in relation to the case study of London Borough of Harrow in Chapter 16

Stephen Gill, formerly of Powys County Council, in relation to the case study of Powys County Council in Chapter 17

At the publishers, we have worked with several individuals. Apart from Tony Moore, our particular thanks go to Sarah Kramer who, during the dark days when progress seemed painfully slow, cajoled and persuaded us not to give up. We would also like to record our appreciation to Matthew Gibbons who picked up the reins when Sarah moved to pastures new. His advice, together with that of Randi Cohen, on finalizing the manuscript and preparing it for publication, has been invaluable.

Finally, our greatest debt is owed to our families, especially our wives Betty and Ann. Without their tolerance and forbearance, there would have been no book at all.

Ian Gilfoyle and Peter Thorpe

Authors

Ian Gilfoyle has an honors degree in geography and is a chartered surveyor and town planner. Until 1998 he was the county planning officer of Cheshire and has 25 years experience using geographic information systems. Ian was a member of the Chorley Committee of Enquiry into the Handling of Geographic Information, a member of Local Government's Geographic Information Advisory Group, and chairman of two of the Ordnance Survey consultative committees. He was a council member of the Association for Geographic Information for 10 years and the convenor of the Royal Town Planning Institute's IT and GIS Panel from 1985 to 2001. Between 1998 and 2001 Ian was an honorary senior research fellow at the University of Manchester.

Peter Thorpe is a geographer and town planner who leads his own consultancy — Peter Thorpe Consulting — that provides IT and GIS advice to town planners, local authorities, and land and property professions. Peter has extensive local government experience gained from working for 20 years in town planning, corporate planning, and IT roles. On leaving local government in 1988, he joined Bull Information Systems as a local government consultant specializing in GIS and land and property issues, before establishing his own consultancy in 1994. Peter Thorpe Consulting has completed over 40 GIS-related projects for local authorities. In 2001, Peter was appointed as the convenor of the Royal Town Planning Institute's IT and GIS Panel.

Contents

Part 3 — The Case Studies

Part 4 — Looking to the Future

List of Boxes and Tables

Boxes

Tables

List of Figures

Abbreviations and Acronyms

AGI	Association for Geographic Information
AM/FM	Automated Mapping and Facilities Management
BCS	British Computer Society
BGS	British Geological Survey
BLPU	Basic Land and Property Unit
BPR	Business Process Reengineering
BSI	British Standards Institute
BSU	Basic Spatial Unit
CAD	Computer-Aided Design
CASA	Centre for Advanced Spatial Analysis
CCTV	Closed Circuit Television
CERCO	Comite European des Responsables de la Cartographie Officielle
CGI	Centre for Geographic Information
CGIS	Canadian Geographic Information System
DBMS	Database Management System
DEFR	Department for Environment, Food and Rural Affairs
DEM	Digital Elevation Model
DETR	Department of the Environment, Transport and the Regions
DNF	Digital National Framework
DOE	Department of the Environment
DTM	Digital Terrain Model
DTLR	Department of Local Government, Transport and Regions
EC	European Commission
ERDF	European Regional Development Fund
ESRI	Environmental Systems Research Institute
EUROGI	European Umbrella Organization for Geographic Information
EUROGISE	European GIS Expansion Project
EUROSCOPE	Efficient Urban Transport Operation of Port Cities in Europe
FVGIS	ForthValley GIS
GI	Geographic Information
GIAG	Geographic Information Advisory Group
GIG	Geographic Information Group
GIM	Geographic Information Management
GINIE	Geographic Information Network in Europe
GIS	Geographic Information System(s)
GISP	General Information Systems for Planning
GLA	Greater London Authority
GLOSNASS	Global Orbiting Navigation Satellite
GML	Geographic Markup Language
GPS	Global Positioning System
GROS	General Register Office for Scotland
HMLR	Her Majesty's Land Registry
HMSO	Her Majesty's Stationery Office

IACS	Integrated Administration and Control System
IAPU	Information Age Practice Unit
ICT	Information and Communications Technology
IDeA	Improvement and Development Agency for Local Government
IEG	Implementing e-Government
IGGI	Interdepartmental Group on Geographic Information
IM	Information Management
INSPIRE	Infrastructure for Spatial Information in Europe
IS	Information Systems
IT	Information Technology
ITN	Integrated Transport Network
ITT	Invitation to Tender
JIS	Joint Information System
JUPITER	Joining Up the Partnerships in the East Midlands
LACSAB	Local Authorities Conditions of Service Advisory Board
LAMIS	Local Authority Management Information System
LAMSAC	Local Authorities' Management Services and Computer Committee
LASER	Local Authority Secure Electoral Register
LGIH	Local Government in Information House
LGIP	Local Government Improvement Programme
LGMB	Local Government Management Board
LGSIG	Local Government Special Interest Group
LGTB	Local Government Training Board
LiDAR	Light Detection and Ranging
LLC	Local Land Charges
LLPG	Local Land and Property Gazetteer
LPG	Land and Property Gazetteer
LSG	Local Street Gazetteer
MAGIC	Multi-Agency Geographic Information for the Countryside
MARS	Merseyside Address Referencing System
MEGRIN	Multipurpose European Ground-Related Information Network
MS	Microsoft
NGDF	National Geospatial Data Framework
NJUG	National Joint Utilities Group
NLIS	National Land Information Service
NLPG	National Land and Property Gazetteer
NLUD	National Land Use Database
NOMIS	National Online Manpower Information System
NRSC	National Remote Sensing Centre
NSG	National Street Gazetteer
NTF	National Transfer Format
ODPM	Office of the Deputy Prime Minister
OGC	Open GIS Consortium (also Open Government Commerce — the new name for the CCTA)
OGIS	Open GIS Specification
ONS	Office for National Statistics

O-O	Object-Orientated
OS	Ordnance Survey
OSCAR	Ordnance Survey Centre Alignment of Roads
OSLO	Ordnance Survey Liaison Officer
PAF	Postcode Address File
PC	Personal Computer
PDL	Previously Developed Land
PGA	Pan-Government Agreement
PITCOM	Parliamentary IT Committee
PLANES	Property, Land and Network System
PLUG	Property and Land User Group
RDBMS	Relational Database Management System
RGS	Royal Geographical Society
RGS-IBG	Royal Geographical Society with the Institute of British Geographers
RICS	Royal Institution of Chartered Surveyors
ROMANSE	Road Management System for Europe
RTPI	Royal Town Planning Institute
SASPAC	Small Area Statistics Package
SCST	Select Committee on Science and Technology
SINES	Spatial Information Enquiry Service
SIMS	School Information Management System
SLA	Service Level Agreement
SOCITIM	Society of IT Managers in Local Government
SQL	Standard Query Language
SUSI	Supply of Unpublished Survey Information
TIIWG	Tradable Information Initiative Workshop
TIN	Trianguated Irregular Network
TITAN	Tactical Implementation of Telematics across Intelligent Networks
TOIDs	Topopgraphic Identifiers
TRAMS	Transport Referencing and Mapping System
UPRN	Unique Property Reference Number
VO	Valuation Office
WAP	Wireless Application Protocol
WMS	Web Map Server
WWW	World Wide Web
XML	Extensible Markup Language

PART **1**

Introduction

CHAPTER 1

The Background to Geographic Information Management in Local Government

KEY QUESTIONS AND ISSUES

- Why a book on geographic information management (GIM) in local government?
- What is geographic information (GI) and how does it relate to spatial data?
- What are geographic information systems (GIS)?
- What are the key features of a typical GIS?
- What distinguishes GIM from GIS?
- Why is GIM important to local government in Britain?
- How has the GIS industry responded to local governments' needs?
- What is the impact of the modernizing government agenda?
- How is the book structured?

1.1 WHY A BOOK ON GEOGRAPHIC INFORMATION MANAGEMENT (GIM) IN LOCAL GOVERNMENT?

After spending well over a year contemplating and researching both the scope and the content of this book, we began writing it as the third millennium dawned, a time described by many as a pivotal point between the past and the future. Early in the year 2000, both the popular and the technical press contained numerous articles by leading writers, commentators, and academics predicting how they saw Britain, the world, and people's lives changing during the 21st century. While their opinions differed, they all recognized that during the past century, science and technology had opened more doors than anyone could have possibly imagined. While science is notoriously unpredictable, most writers foresaw a continuing and rapid growth in information and communication technology. The technical press put more flesh on the bones by detailing the likely developments in Web technology, telecommunications, knowledge-based systems, voice recognition, data availability, and information output. Technological innovations will affect all areas of society, including how people interact with local government and how its services are planned and delivered.

Furthermore, as local authority management becomes ever more complex and accountable, so the role of information as an organizational resource will assume even greater prominence.

Information is central to the workings of local authorities in their day-to-day operations, their management and forward planning, their strategies, and their politics. As Hoggett (1987) neatly puts it: "what poultry is to Kentucky Fried Chicken, information is to local government." Certainly, there is no aspect of the work of a local authority that does not depend on relevant, accurate, and up-to-date information that is available when it is needed and in the form in which it is required. From the electoral register to planning applications, from the housing waiting list to environmental management, from routine road maintenance to personal services, no function of local government can be carried out without information. Council members and staff use information to establish community needs, identify priorities, determine strategies, establish policies, allocate resources, manage assets, and deliver services to meet their aims and objectives.

"Governments run on the bedrock of information. Without current and historical information, readily available and of indisputable quality, the wrong policies can be forged" (Raynsford, 1998). Local authorities have at their fingertips a wealth of information collected as part of their statutory duties. This information is acquired and stored at significant cost, and much of it is geographical in nature in that it relates to some place on the earth's surface. The opportunities opened up by technological change and the government's desire to adapt to the digital age have revolutionized the potential to exploit the vast riches of our geographic information assets.

This book explores the existing practices and possibilities in Britain's local authorities, focusing on the practicalities of the management of geographic information from the user's perspective. Our hope is that it will prove useful to all those interested in local government, whether they are providers or users of services, technical or nontechnical individuals, councilors, officers, or members of the public. We aim to cover the subject matter by raising a series of questions and issues in each chapter and then attempting to answer them, or at least helping the reader to answer them.

1.2 WHAT IS GEOGRAPHIC INFORMATION (GI) AND HOW DOES IT RELATE TO SPATIAL DATA?

GI is derived from spatial data. This is a very broad term that can be legitimately used to refer to datasets as diverse as astronomical observations, topographical maps, and the distribution of stress in a structure or the arrangement of chemical constituents in a sheet of photographic film (Tomlinson, 1986). However, when we use the term spatial data, it invariably refers to geographic data that results from observations or measurements of earth phenomena, particularly those data that describe natural and manmade resources as well as social and economic characteristics. This raw data by itself solves no problems and only becomes information when you ask it questions such as who, what, when, where, and how many.

GI underpins most human activity. It records the location of physical assets like properties, roads, pipes, and cables. It provides an inventory of the natural environment. It supports navigation both personal and collective, and facilitates the optimum location of schools, shops, and hospitals. Finally, it describes the character of an area and the people who live and work within it. Typically, this information may be displayed on a map, or will contain an address, location description, postcode, grid-reference, or Unique Property Reference Number (UPRN). Studies which have been undertaken elsewhere suggest that on average about 80% of a local authority's information is likely to be spatial. For those departments which focus on services at a property, site, or street level (e.g., planning, housing, highways, transport, environmental, and leisure services), this percentage is likely to be even higher. However, much of this information is still held manually and is therefore difficult and time-consuming to access. Even where the data is available digitally, traditional computer systems have tended to "lock in" the geographic aspect so that it is hard to access.

Data is a large complex topic and data capture is expensive. So it is important to ensure that the quality of data you acquire is fit for the purpose for which the data is required. You should always consider whether you require the most accurate data when alternative affordable sources could suit your needs just as well. To help you decide on the appropriate sources, we have devoted Chapter 4 in Part 2 to spatial data.

1.3 WHAT ARE GEOGRAPHIC INFORMATION SYSTEMS (GIS)?

GIS provide the tools to bring together disparate data about the character and activities of a place. They are a representation of reality that answer questions about location, patterns, trends, and conditions, i.e., where features are found, what geographic patterns exist, where changes occur over time, where certain conditions apply, and what the spatial implications of proposed change are. Of necessity, they present a simplified "model" of the real world containing only that data the GIS designer considers necessary to solve a particular problem. However, "GIS are used to help build models where it would be impossible to synthesise the data by any other means" (Martin, 1996). This means that data can be used as a means of coming to grips with systems whose spatial scale or complexity might otherwise put them beyond our mental grasp.

GIS have much to offer local authorities because they allow the spatial dimensions of existing and new information to be exploited, resulting in added value and greater insight. GIS are not a new concept, but simply a new technique. The Domesday Book was, in effect, GIS even though no maps or computers were used, for it listed locations and areas of all land holdings of the Norman Knights with details of the people, major buildings, and animals. The invention of topographical maps greatly extended the capacity of GIS. By the use of symbols, overlays, and diagrams, these maps soon began to contain a wealth of other information. More recently, the use of new technologies and modern computers has dramatically changed and extended the way in which GI is handled.

Computerized GIS have been used since the 1960s to help make decisions. However in those days, the field of GIS was seen as a science administered by highly technical people using powerful computers and specialized programming languages that gave their results in paper form. Vanessa Lawrence (1998) explains that "today, the world has changed; GIS has become part of the desktop suite of business applications and now it is possible for everyone in an office or mobile environment to have GIS on their desktop. It is now being used by people with no specialist or geographical training to make critical business decisions using a new dimension which until recently was not considered of any importance to them; that being the spatial location or geographical dimension."

Even among academic writers there is still disagreement about what exactly constitutes GIS and what functions they should perform. This is partly because they have grown out of a number of technologies — computer-assisted mapping and design, remote sensing, digital mapping, database management, image processing — and a variety of applications, and people see GIS from their own particular viewpoint. Above all else, GIS are integrating technologies. They bring together these once-separate technologies, thereby enabling the handling of geographic data in ways that were not previously possible, as well as allowing much more sophisticated spatial analysis. In addition, GIS are integrating technologies in terms of their abilities to pull together all that is available about place in terms of data.

Of the many definitions of GIS, we prefer to use the definition from the authoritative *Report of the Committee of Inquiry into the Handling of Geographic Information* (the Chorley Report), as this was the result of a meeting of minds from many disciplines and backgrounds over a 2-year period. This report describes a GIS as a "system for capturing, storing, checking, integrating, manipulating, analyzing, and displaying data which is spatially referenced to the earth" (DOE, 1987).

1.4 WHAT ARE THE KEY FEATURES OF A TYPICAL GIS?

The key features of a typical GIS are shown in Figure 1.1. These enable a local authority to:

- Take full advantage of the Ordnance Survey's digital base map of the local authority area and the service-level agreement by which it is kept up-to-date through the receipt of regular replacement maps recording changes (❶ and ❷).
- Hold "user overlays" of information relevant to the needs of different departments (e.g., land charges searches, listed buildings, planning application boundaries, council-owned properties, wards) as layers above the basemap (❸).
- Produce digital maps for different users to meet a wide variety of requirements (varying size, scale, notation, and joined seamlessly across traditional map boundaries) (❹).
- Create links between the digital map and data held in external databases (e.g., census) or processing systems (e.g., planning application processing system, environmental health system, housing system) (❺).

It is the last item ❺ that distinguishes GIS from the ability merely to hold, manage, and display computerized maps, and it is where the real "power" of GIS technology lies. A key requirement of all local authorities is the ability to pull together information that is known about a specific plot of land or property from different depart-

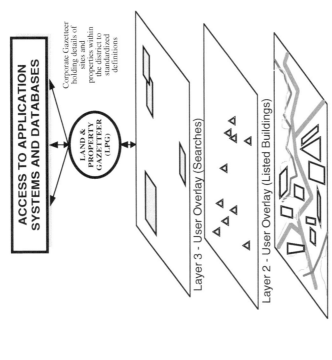

ACCESS TO APPLICATION SYSTEMS AND DATABASES

LAND & PROPERTY GAZETTEER (LPG)

Corporate Gazetteer holding details of sites and properties within the district to standardized definitions

Layer 3 - User Overlay (Searches)

Layer 2 - User Overlay (Listed Buildings)

Layer 1 - Basemap

❶ Digital basemap of local authority district (from Ordnance Survey)

❷ Updating of basemap (through OS Service Level Agreement)

❸ User overlays held as layers above the basemap (many overlays of corporate interest will be built up for local land charges purposes)

❹ Digital maps plotted to meet widely varying requirements

❺ Links (in both directions) to external application systems and databases

❻ A corporate Land and Property Gazetteer is a fundamental requirement if unambiguous interchange of data between departments is to be achieved

NOTES:
Digital Mapping covers ❶ ❷ ❸ *and* ❹ *above. It is the ability to create links between the map and data (held in application systems/databases) which distinguishes true GIS (* ❺ *above).*

Figure 1.1   Key features of a GIS.

ments and systems. Use of GIS in order to support the integration of information in this way is a good example of how sharing of information across council services and the resulting improvements in cross-team working can be facilitated.

A prerequisite for the improved ability to integrate information between departments is the adoption of corporate data standards. This means that an authority would need to consider introducing standard "data building blocks" (e.g., for streets, properties, and sites) if the bringing together of information from different computer systems is to be readily achieved. Without the adoption of corporate data standards, each department would continue to hold data relating to land and property by different forms of address, descriptions, and reference numbers, making interchange of information difficult. An increasingly important way of putting this into practice is through the implementation of a corporate Land and Property Gazetteer (i.e., a master list of properties and sites within the local authority area) that conforms to the British Standard BS7666. (See also Sections 7.3 and 7.5.) As shown in Figure 1.1, establishing a corporate Land and Property Gazetteer (❻) is a fundamental requirement especially if unambiguous interchange of data between departments is to be achieved. While this gazetteer could be implemented initially on a stand-alone basis, for example, to support a local land charges computer system, there is the opportunity for a local authority to extend its use eventually if appropriate to all computer systems that hold land and property information.

As we will demonstrate later in the book, a wide variety of GIS are being used for a great diversity of purposes. So, systems for handling GI come in many different forms, and it is important to recognize that an off-the-shelf product may not solve your problems and that successful implementation will always require considerable thought before launching into expensive hardware and software.

1.5 WHAT DISTINGUISHES GIM FROM GIS?

It is important to distinguish between GIS and GIM. GIM deals with the whole process associated with the development and operation of computer-based systems for geographic information handling to satisfy the needs of specific users. It takes into account organizational factors as well as technical issues.

At the risk of over-simplification it is the difference between a set of tools, that is a GIS, and the various conditions which have to be satisfied if these tools are to be used effectively or, for some applications, whether these tools are to be used at all, or whether (to depart from the tool-kit analogy) they give meaningful answers (Chorley, 1991).

Against this background, Box 1.1 contains our definition of GIM.

Box 1.1   Geographic Information Management

GIM covers all the activities involved in planning, promoting, and administering geographic information as a key local authority asset that directly support its corporate and service priorities. This includes encouraging the responsible ownership of geographic information, facilitating its supply, and ensuring its currency, quality, and accessibility.

1.6 WHY IS GIM IMPORTANT TO LOCAL GOVERNMENT IN BRITAIN?

The issues covered in this book, whether about methodologies or applications, focus on the aspirations, needs, and capabilities of those in local authorities who both manage and use geographic information in the delivery of essential services. Therefore this section contains a brief overview of local government in Britain at the start of the third millennium, indicating the importance of GIM.

For years local government has been on the frontline of providing essential services that meet ever-rising quality standards within ever-tightening financial constraints. Traditionally, local authorities are in business to provide a very wide and varied range of services for people living and working in, or visiting, their area. The scale and complexity of this service provision is reflected in the fact that individual local authorities are often one of the largest employers in their area, spend in total over £75 billion a year from taxes, and undertake some 700 different activities. Despite the 16% reduction in numbers achieved by the local government reorganizations between 1996 and 1998, there are still 441 authorities in Great Britain. Nearly 40% of these are unitary authorities, including the London and Metropolitan Boroughs in England, the remainder of the country being divided into 34 counties containing some 238 district councils.

All these authorities have their own responsibilities, needs, and aspirations, their own culture, styles of leadership, and their own approaches to decision making. They take pride in the services they provide and are constantly seeking improvements to both their efficiency and effectiveness.

For many years, local government has been under continuous pressure to change. It has proved itself to be more dynamic and flexible than many people thought possible, readily accommodating some fundamental restructuring arising from new legislation, becoming enabling authorities, increasingly competitive, and devolving control both internally and externally. "We have also seen the emergence of more effective and coordinated 'top down' executive leadership, producing clear mission statements, corporate plans, service delivery plans, and a plethora of charters and promises to the clients, backed up by improved performance and better information services" (Gill, 1996). Over the last 10 to 15 years, local government has been required to become more businesslike and commercial with the introduction of resource center plans, cost centers, service level agreements, and internal charging. As a result the culture of most authorities now positively encourages change where it can be shown to be beneficial.

Local authorities are essentially "people" organizations in that they serve people by employing staff. In fact, staff are now recognized as one of their greatest assets, a resource to be developed and managed with care if their potential is to be realized. These matters are being addressed by the introduction of new management practices such as performance management, quality assurance, more personal accountability, and investors in people. The continuing development of information technology (IT) in local government relies as much on the skills and knowledge of its staff as on adopting best practice.

The Local Government Act 2000 sets out legislation to allow local authorities to "promote the well-being of their communities" by entering into new forms of partnerships with each other, with other statutory service providers, and with both the voluntary and private sectors. The new act, the drive for best value, changes to local government finance, and e-government are all part of the radical agenda designed to change local authorities from service providers to enablers. Partnership among the many agencies providing services is increasingly seen as an effective way of tackling both individual and community issues. Closer work among local authorities, health providers, police, voluntary organizations, education and training providers, and local commerce and industry is becoming the accepted way forward on issues such as drugs misuse, crime and disorder, the environment, poverty, social exclusion, the needs of young people, the unemployed, and the frail elderly. GI is essential both to understanding and addressing these issues.

All these changes give more power to people, place more emphasis on account-ability, and increase the popularity of "one-stop shops" where the local community can gain access to information about a whole range of local government and other services in their area.

Thus, in the early years of the new millennium, local authorities face changes as radical and as rapid as at any time in the past. There are changes in the needs of their citizens, constant changes in the legislation affecting their key services — education, housing, social services, environment, transport, recreation, and leisure — and changes in the way that they are financed. The pace of change in the world around us shows no signs of slowing; in fact, it is accelerating. Society is changing as people become more aware of their power as consumers, demanding even more from those organizations that supply them with goods and services. Local politicians need to know more about their constituents and have an accurate up-to-date profile of the area they represent. They also need information for setting priorities, allocating resources, and assessing performance.

In order to meet all these information needs, many authorities have prepared corporate information management (IM) strategies that lay down common values, principles, and requirements and set out how the information needs of the organi-zation will be met.

1.7 HOW HAS THE GIS INDUSTRY RESPONDED TO LOCAL GOVERNMENTS' NEEDS?

The GIS industry has responded well to the needs of users and the aspirations of governments. During the second half of the 1990s, Autodesk, Microsoft, and Cadcorp joined companies like ESRI, Intergraph, and Laserscan that were estab-lished as GIS suppliers over 30 years ago. When GIS were first introduced, the systems were proprietary, large mainframe systems, the programming languages were highly specialized, the data could not be transferred to other systems, and the systems did not integrate with the corporate IT strategies. Access to spatial data was often prohibitively expensive and data file sizes frequently too large to store.

Over the years the increased power of hardware, the phenomenal growth of Microsoft, new data suppliers, and user-friendly software have changed the GIS industry from a highly specialized niche technology to a part of a desktop suite of business applications. Many customers now see the value of using an added geographic dimension for decision making, and this growing user demand has given impetus to the development of the Open GIS Consortium. This consortium is a group of GIS developers, academics, and government organizations set up to define and encourage open system standards for the GIS industry. Its main objective is to facilitate interoperability between GIS software and data formats.

1.8 WHAT IS THE IMPACT OF THE MODERNIZING GOVERNMENT AGENDA?

Governments, both central and local, are not only the biggest suppliers of geographic information; they are also the biggest users. So, they stand to be the biggest beneficiaries of an enlightened approach to data management. If their senior managers and politicians have the vision, then both central and local governments stand on the edge of a geographic information revolution.

The modernizing government agenda, with its strategy for information age government, is an impressive start. It focuses on better services for citizens and businesses together with the more effective use of the government's information resources. The e-government strategy challenges all public sector organizations to innovate and contains within its guiding principles the desire to treat the government's own information as a valuable resource. While the government recognizes the importance of face-to-face contact in many services, it wishes to achieve the rapid development of electronic service delivery wherever possible. This is part of its vision to build service delivery around customers' needs, not around the organizations of government.

Great Britain's prime minister has issued a challenging target proposing that all public dealings with local government should be capable of electronic delivery by 2005. The e-government vision requires a massive culture change on the part of local government. The white paper *Implementing e-Government: Guidelines for Local Government* (Cabinet Office, 2000a) proposes that each council should set its own electronic service delivery targets, as far as practicable consistent with that goal, linking these targets to the achievement of best value. *Information Age Government: Targets for Local Government* (DETR, 2000) propounds that such targets should address two specific themes: corporate approaches to managing information and service specific targets. We examine both of these themes in the Part 2 of this book.

1.9 HOW IS THE BOOK STRUCTURED?

The book is organized into 4 parts and 18 chapters. It is structured in a way that enables you either to browse through the text or to use the book as a reference source, reading it chapter by chapter. For those with limited knowledge of GIS we recommend reading each chapter in sequence in order to discover the ideas presented and develop an understanding of the subject at your own pace. The next chapter traces the development of GIM in local government and completes the introductory part of the book. Part 2 and Part 3 comprise the bulk of the book. The first of these deals with the key elements of GIM. It has often been said that a GIS is like a three-legged stool, the three legs being software, data, and those services that aid implementation. If one of those legs is loose or unstable, the stool could topple over and become totally unusable — even dangerous. (Kendall, 1999).

To take this analogy a bit further, just as the stool has to be designed to carry the weight of all those who will sit on it, so it is vital to understand how the legs of GIS will support GI needs and aspirations. No GIS should exist in isolation from its organizational context. There must be people to plan, implement, and operate the system, as well as to make decisions based on the output.

Many GIS implementations go wrong because organizations are not sure how to use them. Therefore, we start the second part of the book by considering the human and organizational issues in Chapter 3. Chapter 4, Chapter 5, and Chapter 6 concentrate on the three main elements or legs of GIS, covering data (the raw material), the technology toolbox (particularly the software), and GIS justification, selection, and implementation. GIS operate in a fast-moving world of new ideas and opportunities, so it is important not to reinvent the wheel. To assist in the exchange of ideas, to develop standards, and to disseminate good practice, a number of coordinating mechanisms are now in place, and these are discussed in the final chapter of Part 2.

Part 3 of the book is given over to an in-depth analysis of nine case studies. These draw heavily on the material supplied by lead contacts in response to our questionnaire, which is reproduced in Appendix 1. They have been compiled to bring out different approaches using a standard template. To aid comparison, each of the case studies is summarized using a standardized bullet-point format.

Part 4, the final part, is forward looking. Here, with the help of Michael Batty, we assess the prospects and challenges for GIM in local government.

CHAPTER 2

The Development of GIS in Local Government

KEY QUESTIONS AND ISSUES

- What methods were used for handling spatial data before GIS?
- Who were the early pioneers of GIS?
- What was the Chorley report and what was its impact?
- To what extent did the Chorley report lead to the coordinated development of GIM within local government?
- How has GIS spread through local government during the 1990s?
- What progress had been made ten years after the Chorley report?
- How has the modernizing government agenda affected GIS development?
- What lessons can be learned from this review of GIS development?

2.1 WHAT METHODS WERE USED FOR HANDLING SPATIAL DATA BEFORE GIS?

Before computers were commonly available in local authorities, maps were the main means of handling GI. They were used in a whole range of applications, for example, plotting planning applications, recording property terriers, route finding, calculating areas, locating new school sites, and understanding patterns of crime and accidents. In fact, maps were so important that a wide variety of professions working within local government could not exist without them. Sieve mapping, a technique combining several map layers on a light table to identify areas of overlay and interest, was particularly important in handling spatial data. But paper maps have their own, well-known limitations; they stretch and shrink, they are always out of date, they often require complete redrawing for minor changes, they need considerable storage space, they can be easily lost or destroyed, and they literally fall apart if in constant use for years.

For those interested in environmental and land use information, aerial photographs have been valuable sources for many years. While they provided "snapshots" of geographic areas at particular instants in time, they were expensive to obtain, affected by cloud cover, and, because of poor ground resolution, the depicted details were often too coarse for most local government activities. However, because they involve repeat photography of the same area, aerial photographs remain particularly useful as background information and for monitoring environmental and land cover change. For this reason, many local authorities had their areas surveyed by air to coincide as closely as possible with the decennial population census, thereby providing a physical background to the socioeconomic information.

Other spatial data, including population census records, unemployment figures, environmental health complaints, and pupil and client records, were dispersed through files, books, or microfiche and so were even more difficult to access, integrate, and compare. In reality, the sheer volume of data and difficulty of manipulating it by hand precluded much-needed analysis.

Given all these problems, it was hardly surprising that those personnel involved in land, property, and transport matters — highway engineers, surveyors, planners, and valuers — were keen to explore computer technology even though initially it was expensive and time consuming.

2.2 WHO WERE THE EARLY PIONEERS OF GIS?

In Britain the first application of computer technology to the handling of GI occurred in the late 1960s with the systems approach in subregional planning studies such as Nottingham and Derbyshire, and Coventry, Solihull, and Warwickshire, and in land use and transportation studies such as Merseyside. The main rationale was that all human activity was a system and that, given enough data, the computer could model these systems, predict how they would change in the future, and then produce alternative development proposals.

This approach faltered because it failed to recognize that the real world is a system of such complexity that it could not possibly be modeled by something so crude as mathematical formulae fed into a computer. In addition, it overlooked that many decisions affecting people's lives are made within a political context, and so the decision-making process itself is subject to unpredictability (Allinson, 1994). However, the early pioneers learned that, even if the computer did not have the capability to make actual policy decisions, its power was in its capacity to hold, manipulate, and make available large quantities of information, thereby giving support both to operational and policy decision making.

The GIS pioneering effort began in the 1960s. One of the earliest, if not the earliest, examples of GIS resulted from the creation of the Canada Geographic Information System (CGIS) under the direction of Dr. Roger Tomlinson, known by many as the "father of GIS." His role was to process the immense amount of data created by the Canada Land Inventory. In the sixties, the Canadian government felt, perhaps for the first time, that although its natural resources were extensive, they were not limitless. A special committee of the Senate was established to examine land use in Canada and a nationwide land inventory was initiated. While Canada could afford to gather the data and make the maps, the manual techniques of map analysis required thereafter were extremely labor intensive and time consuming. CGIS's trail-blazing project was well ahead of its time in digitizing map data, edge-matching map sheets, and developing a spatial database management system with descriptive attribute information for each zone. It also had the advantage of being politically motivated.

Shortly afterward, a number of other systems began in North America, including those within the states of New York, Minnesota, and Maryland and the municipality of Burnaby in British Columbia, Canada. These systems emphasized facilities information. However, this was not restricted to the physical infrastructure but included land-use zoning, traffic accident details, crime statistics, population distribution, property ownership, and much more. Their digital records were referred to as AM/FM, an abbreviation of Automated Mapping and Facilities Management. This was broadly defined as computer-aided cartography (AM) and the management of the business or information that can be made from records that are associated with the map (FM). The AM/FM methodology spread to Europe and was used extensively by public utilities in the U.K.

Back in the U.K., the Local Authority Management Information System (LAMIS) was developed in the early 1970s by International Computers Limited (ICL) and implemented at Leeds. It was originally intended to satisfy the requirements of local land charges by defining all property boundaries in spatial terms and producing an integrated set of data files serving the corporate needs of the council. ICL moved on to develop the LAMIS approach on its more advanced 2900 range of computers between 1977 and 1978 for Dudley Metropolitan Borough Council. This online property system was distinguishable from previous versions by the use of relational databases rather than serial files for storing data.

Several other local authorities, including Brent, Doncaster, Birmingham, and South Oxfordshire, developed LAMIS and LAMIS-type systems to support central property registers, planning application processing, land potential and development monitoring, and housing information services. "Huge paper listings were the order of the day, every property had a Unique Property Reference Number (UPRN), but the system was inflexible and difficult to use and ever-rising operating costs led to its gradual demise" (Humphries and Marlow, 1995).

A DOE report in 1972 titled "General Information Systems for Planning" (GISP) was an early attempt to tackle the general problem of how to organize information, including the question of geographic referencing. GISP (1972) originated as a means of organizing planning-related data, but it was quickly realized that the information needs of the planning department could not be separated from those of the remainder of local government. GISP recommended the use of common basic spatial units (BSUs), each of them with a UPRN.

However, for GISP to have worked efficiently at all levels of local and central government, a single national gazetteer would need to have been implemented, together with a common BSU framework applied to all data. As a result, in 1973, the National Gazetteer Pilot Study was funded to investigate the problems of gazetteer creation in the multilocal authority Tyne and Wear County in preparation for the 1981 population census. In 1975, the system was renamed the Tyne and Wear Joint Information System (JIS) although the first reliable gazetteer was not available until 1978. This covered 5 metropolitan districts and consisted of 500,000 BSUs each with a UPRN, a land-use code, a 1 meter grid-reference, a postal address, and

Box 2.1   Merseyside Address Referencing System (MARS)

MARS is essentially a computerized directory of streets and addresses based on a digitized version of the County's road network. It uses segments of the road network as the basic structure; these are accurately positioned by recording the OS grid co-ordinates of the two end nodes to one metre resolution. Properties can be associated with segments of the road network since all properties have access to the road network and all postal addresses combine their address number or name with the street name.

Address gazetteers are an important product of MARS and represent the main requirement of its primary sponsor, the Police, who use it for their Command and Control System. The gazetteer facility can provide comprehensive lists of addresses for particular areas. A Street Index containing records for all potential incident locations is used for operational control and subsequent analysis of incident patterns.

Source: From the DOE (1987) *Handling Geographic Information: Report of the Committee of Enquiry,* London: HMSO.

a few area indicators. Writing in the Chorley report, Stan Openshaw concluded that "JIS certainly indicates that GISP can be made to work and that the real advantages are only likely to become significant when local authority-wide corporate databases can be established and run for a number of years" (Openshaw, 1987). In support of this conclusion, it is interesting to note that the Joint Gazetteer team continued to be funded until the early 1990s, and while the team has now been disbanded, the components of the gazetteer have generally been adopted as the basis for further development by the original local authorities.

The principal difference between gazetteer systems and LAMIS was that the central indexing was based on postal addresses, not on spatially defined areas. Each property record had a single spatial reference in the form of a grid reference of the property centroid rather than digitized boundaries.

Merseyside adopted a different approach to property referencing (see Box 2.1). Here the Merseyside Address Referencing System (MARS) was a network-based system developed originally by the Transport and Road Research Laboratory to provide a common locational referencing system for highway data — the Transport Referencing and Mapping System (TRAMS).

Apart from being used for command and control, MARS played a central role in defining priority areas for funding by enabling different sets of deprivation criteria to be displayed in map form. Identifying areas of family stress was also one of the earliest applications of the use of GIS in neighboring Cheshire, where 16 indicators of stress — such as the number of free school meals, probation orders, and youth employment rates — were collated for each ward across the county. Maps shaded by ward were plotted for each indicator and a summary map showing the 10 most heavily stressed areas was produced. (Gilfoyle and Challen, 1986). This improved the understanding of the varying needs throughout Cheshire both by councilors and senior officers as well as contributing to the more efficient allocation of resources by focusing them on the priority areas.

Early in the 1970s, the Ordnance Survey (OS) had recognized that the bedrock of any GIS in Britain would be its topographical database in digital format. So, in 1973, it began a nationwide program of digitizing all of its 230,000 base scale maps. Initially, progress was slow, and by the middle of the 1980s only 13% of the map sheets were available for purchase on computer tape. At that stage the OS expected to complete the digitizing of the major urban areas by 1995 and the rest of the country by 2005. Most local authorities took the view that this important digital database would be of little value to them until the majority, and preferably the whole, of their area was completed. Therefore, for many years the OS digitizing program made little impact on most local authorities.

Dudley, which had used the LAMIS approach to develop a computerized land and property system, was one area where there was early coverage of OS digital mapping. So the Dudley Council was able to provide an early example of a corporate approach to the use of geographic information. Its aim was to produce an integrated set of data files serving the corporate needs of the council and thereby achieve greater efficiency in the use of resources. The initial uses of the database were in planning applications, building inspections, general enquiries, and a property terrier. However, it later developed into a unique databank able to serve the needs of all the utilities as well as the local authority. This Dudley digital records trial was also an early example of GIS development driven by a business case. A survey conducted by the National Joint Utilities Group (NJUG) in 1981 had indicated that the national cost of mains and plant damage due to one utility digging up another utility's pipes and cables was of the order of £14 million per annum. It was the recognition of the opportunity for substantially reducing this amount by exchanging and sharing the same digital data that led to the successful 5-year Dudley trial.

A corporate property database was also the aim of the City of Glasgow District Council. In 1979 the council decided to invest in a computer-based system primarily to manage its housing estate, then the largest in Europe. The council subsequently arranged with the OS for the provision of digital map data for the city, which could be supplemented with the council's operational and administrative boundaries as well as its property database to enable these attributes to be associated with the maps.

The use of remotely sensed data from both airborne and satellite sensors grew in importance during the 1970s and the early 1980s with the launch of the USA Landsat and the French SPOT remote sensing satellites. While the growth of satellite imagery greatly increased the capacity for the monitoring and management of the environment, it was comparatively little used in local government. This was because of the poor ground resolutions achieved by the earlier satellites, together with the high costs involved both in acquiring the data and in the image processing necessary to produce usable results.

The 1980s saw the emergence of the personal computer (PC), digital maps, and package software. Some online systems appeared, but most were still based on large processors and there was a preoccupation with implementing and using them with rather less emphasis on housekeeping and data management (Humphries and Marlow, 1995). However, it was no longer necessary to seal away the computer in some hallowed place only to be touched by the high priests of information technology; IBM developed the IBM PC, a machine that was completely self-contained and robust enough to sit on a desk in an office or at home (Allinson, 1994). The user became the operator, often networked with others, and data was entered directly via a keyboard rather than by punched cards. In local government, administrative efficiency was very much the keynote of the times, and the PC revolution came just in time to enable local authorities to rise to the challenges of wastefulness and bureaucracy. However, while the early PCs were very good at repetitive, routine tasks for which they possessed all information, they lacked the memory size and graphics capability to handle the more sophisticated mapping and geographic information management (GIM) roles (Allinson, 1994).

In 1986, the English metropolitan counties were abolished and the significant transfer of strategic powers, information, and computer systems to the unitary metropolitan districts "became something of a badly organized lottery" (Davies, 1995). Some systems, such as MARS, were retained intact, but Davies believes that the major strategic importance of data was lost. "On abolition, many of the key personnel involved in centrally controlled systems either retired, left for other jobs, or changed their own career paths. Their expertise was lost as was much of the meta-data" (Davies, 1995). The importance of geographic information and corporate data sharing had not been widely recognized in the process, which was particularly unfortunate because the information had been collected at public expense.

2.3 WHAT WAS THE CHORLEY REPORT AND WHAT WAS ITS IMPACT?

It was therefore opportune that the *Report of the Chorley Committee of Enquiry into the Handling of Geographic Information,* published in April 1987, raised the whole profile of IM in local government. This committee was appointed in April 1985 with the following terms of reference: "To advise the Secretary of State for the Environment within two years on the *the future handling of geographic information* in the United Kingdom, taking account of modern developments in *information technology* and of *market need"* (DOE, 1987, author's emphasis).

The report built on the work of two earlier committees: the OS Review Committee chaired by Sir David Serpell (Serpell, 1979) and the House of Lords Select Committee on Science and Technology on "Remote Sensing and Digital Mapping" chaired by Lord Shackleton (SCST, 1983). The first of these reports was wholly concerned with topographic mapping with only one chapter (out of ten) devoted to the idea of holding maps in computer form. The second report took account of the rapid change in computer technology that had occurred in the intervening period and argued that the computer potentially enabled an enormous range of previously disparate data collected by government to be collated and related together (Rhind and Mounsey, 1989). It was this report that recommended the appointment of the Chorley Committee of Enquiry.

The Chorley Committee received written evidence from 400 organizations and individuals (of which 144 were local authorities), oral evidence from 26 organizations, and a range of presentations and reports on specific topics. In addition it held 22 meetings and undertook 30 visits. The breadth of this review gave a clear picture of the contemporary use of GI, showed that the availability of the technology itself, however necessary, was not a sufficient condition for its effective utilization, and allowed the committee a unique insight into the barriers that were likely to impede progress. The evidence of Roger Tomlinson, previously referred to as the father of GIS, stressed the point that "the success or failure of GIS rarely depended on technical factors, almost always on the human and managerial ones," (Tomlinson,

1986) and led the committee to focus attention on the human and organizational barriers to the uptake of the technology rather than the technology itself.

Reviewing his committee's report in the opening address to an AGI symposium held 10 years after publication, Lord Chorley recognized that it was this user and organization approach that has given many of the report's findings "a certain time-lessness." The basic issue remained, that is, "can human attitudes and perceptions, and their creators — organizations — keep up with the speed of technical progress?" (Chorley quoted in Heywood, 1997).

According to Heywood (1997), the committee identified the following needs:

- Complete national coverage of the OS's large-scale digital maps
- Increased awareness of what technology and data were available and what would be possible in the future
- Access to the vast amounts of spatially related data held in a wide range of government, business, and voluntary organizations
- An enhanced program of research and development that would see a partnership between industry and academia
- A credible cost–benefit methodology that would allow businesses to establish a case for the use of the technology
- The establishment of a national center to coordinate and promote GI activities

The Chorley Committee took a refreshingly broad view of GIM and highlighted the potential to use GIS for environmental monitoring, rural resource management, epidemiology, network management, vehicle navigation systems, marketing and business location as well as the more traditional planning, engineering, and property roles.

The formal government response to the committee's 64 recommendations was published in a March 1988 document that was bland by the standards of the government responses to the preceding Serpell and House of Lords reports (Rhind and Mounsey, 1989).

The Chorley report led to an acceleration in the OS's digital mapping program, prompted a Tradable Information Initiative within central government, and generally increased awareness of GIS potential. All this despite the fact that it did not result in the desired response from government for either open access to government-held information or the setting up of a national center for GI. The recommendations of the committee, coupled with the government's less than enthusiastic response, were perhaps the catalysts needed to stimulate the GI community to coordinate its activities and establish an independent GI organization, the Association for Geographic Information (AGI). The result was a healthy, though perhaps inward-looking, community independent of government (Heywood, 1997).

2.4 TO WHAT EXTENT DID THE CHORLEY REPORT LEAD TO THE COORDINATED DEVELOPMENT OF GIM WITHIN LOCAL GOVERNMENT?

The AGI was founded in 1989 as a multidisciplinary body dedicated to advancing the use of geographically related information. Its aims were to increase awareness of the benefits brought by the new technology and to assist its practitioners. The AGI continues to promote the technology through holding annual conferences and exhibitions, organizing interest groups, encouraging GIS research, collecting and disseminating GIS information through its publications, as well as developing policy advice. Since its formation, the AGI has had considerable success in its task of informing, influencing, and acting in the interests of the GI sector, and has produced a number of reports. Those of particular relevance to local government include a review of GIS in schools written by Diana Freeman (AGI, 1993) and guidelines for GI content and quality (AGI, 1996).

The Local Government Management Board (LGMB) was formed in April 1991 by a merger of the Local Government Training Board (LGTB) and the Local Authorities Conditions of Service Advisory Board (LACSAB). The LGMB was governed by a board of elected members nominated by the local authorities associations. It represented the interests of all local authorities throughout the country.

By focusing on management and human resource issues, the LGMB helped local authorities more effectively deliver services and provide democratic leadership in their communities. The LGMB was supported by a number of advisory groups including the Geographic Information Advisory Group (GIAG), which took over from the earlier Geographic Information Steering Group in July 1991. GIAG had a membership comprising representatives from local authorities, the local authority associations, and the LGMB, and a remit to advise local government, promote good practice, and protect its interests in the field of geographic information. A period of frenetic activity followed through the establishment of working parties, the commissioning of consultants' reports, and the publication of a whole range of influential reports.

In 1988, an Audit Commission report, *Local Authority Property — A Management Overview,* summarized research that estimated that around 20% of the cost of local authority services could be related to property. But it also concluded that property was often regarded as a "hidden resource" over which individual service departments have little control (Audit Commission, 1988). So it is not surprising that land and property became a particular focus of LGMB activity leading to the establishment of a specification for a National Land and Property Gazetteer as well as a National Street Gazetteer. These provided an opportunity for some of the earlier ideas that started off in LAMIS and GISP to be resurrected.

While these are the best known and most widely used of the LGMB's documents, there were numerous others, including reports giving guidelines or methodologies for benchmarking, evaluating GIS, assessing spatial data quality ("fitness for purpose"), functional specification, and cost-benefit analysis. There were also reports relating to experiences in GIM and case studies relating to planning in Welwyn, education in Bedfordshire, and social services in East Sussex. The Welwyn experience demonstrated the laudable aim of providing a seamless mapbase for the authority, but also highlighted the time and resources required to capture the operational information from just one department (LGMB, 1991; Ball and Simmons, 1993). The social services case study contains some important messages for those local government professionals who are not traditional users of map-based information. These are summarized in Box 2.2. Through its Good Practice Working Party and the "Go with the Flow" Initiative (LGMB, 1995), GIAG promoted the advan-

Box 2.2   GIS in Social Services — Eight Key Messages

1. The ability of social services departments to use information more effectively to plan and deliver services was highlighted by the Children's Act 1989 and the NHS and Community Care Act 1990.
2. The requirement to identify children in need and provide appropriate support services will involve the use of a range of information from various sources.
3. Why is GIS important? With the increasing need to present and support policy decisions there is a need to find a new and improved language of communication amongst the service enablers and providers.
4. The planning and regulatory role of social services and the associated need to procure services makes efficient planning and monitoring become essential activities. GIS can be used to support these activities.
5. GIS is about improving the quality of service delivery, it is not about technology.
6. GIS is a strategic decision making support tool and its use should be supported by senior management.
7. The project should be end user defined and driven, and not interpreted by the software developers.
8. Finally, as long as relevant social services data incorporates a geographical reference there is no necessity for social services personnel to acquire GIS skills. The GIS services could easily be purchased through an inter department service agreement, thereby maximising and adding value to existing information without high technical or financial overhead.

Source: Extracted from Local Government Management Board (1994) *Information for Caring: GIS in Social Services,* Luton: LGMB.

tages of the use of structured methodologies, emphasizing the analysis of data holdings, flows, and processes.

Perhaps the most important role of GIAG was negotiating the first Service Level Agreement (SLA) between the local authority associations and the OS, which came into effect on April 1, 1993. This provided OS with guaranteed revenue income for 3 years in exchange for a copyright license package that effectively meant that all local authorities could use OS digital data and graphics products in connection with all their service activities and statutory duties in exchange for the payment of the negotiated annual service charge. The SLA has subsequently been revised and extended in scope.

In addition to the OS's topographical database, the Chorley Committee recognized that central government holds vast amounts of spatial data that is of value to local authorities and others. This led to a Tradable Information Initiative Working Group (TIIWG) being established in 1989, comprising all government departments using or holding geographic data. In conjunction with the AGI, a metadatabase was established to store all the data and to make it generally available. In 1993 the Inter-Departmental Group on Geographic Information (IGGI) succeeded the TIIWG with the aim of facilitating the effective use geographic data both within and outside government. Considerable progress was made following a series of roundtable discussions in 1995 when the AGI, in conjunction with the IGGI, examined the barriers to the wider use of government GI. The most important initiatives that emerged from these roundtable discussions were the National Geospatial Data Framework (NGDF) and the GI Charter Standard Statement, first produced in 1997 (Heywood, 1997).

Improving services to the citizen by public authorities had been the central theme when the government promoted the Citizen's Charter in 1992. It was in the housing section of that charter that the government first expressed its commitment to the development of a National Land Information Service (NLIS). This, in turn, stimulated the computerization and accessibility of the Land Registry, which had been opened to public inspection in December 1990. It also provided a boost for coordinated working among the LGMB, the OS, the Valuation office, and the Land Registrary (HMLR), as well as for other "N-initiatives," which we will explore in Chapter 7.

Further coordination was taking place among the GI associations in Europe, and in 1993 EUROGI (the European Umbrella Organization for Geographic Information) was formed with the mission to promote, stimulate, and support the development and use of geographic information and technology throughout the continent. It became an association of associations playing a major role in the definition of the GI2000 initiative of the European Commission concerned with the supply of European GI that is easily identifiable and accessible; when this was abandoned, it developed its own strategy. However, EUROGI has so far had little impact on U.K. local government.

In contrast, stimulated by pressure from users and suppliers, the Open GIS Consortium (OGC) that was founded in 1994 to address the issues of interoperability, data access, standards, and specifications has had an important, if indirect, influence on GIM in local government. This global consortium of representatives of the GIS industry, utilities, government agencies, and academics has done much to remove the restrictions to the widespread adoption of GI by individual users, vendors, and the wider community. OGC's approach has been to build consensus and promote a vision by creating an Open GIS Specification so that discrete GIS software components can communicate with each other using specified, open interfaces. In April 1998 the OS became the first non-U.S.-government agency to add its name to the growing membership of the OGC.

2.5 HOW HAVE GIS SPREAD THROUGH LOCAL GOVERNMENT DURING THE 1990s?

The early 1990s saw widespread diffusion of GIS in local authorities. This is illustrated by four surveys. The first, undertaken in 1991 by Sheffield University, provided a benchmark dataset. This was followed in 1993 by a telephone survey of all 514 local authorities carried out by Ian Masser and Heather Campbell at Sheffield University in conjunction with LGMB's GIAG. Table 2.1 gives an overall picture of the state of local authority plans for GIS at the time of both surveys. From this it can be seen that in the summer of 1993, 29% of all local authorities in Great Britain had GIS facilities compared with 16.5% in 1991. In addition, a further 10% had firm plans to acquire GIS facilities within a year (Masser and Campbell, 1994).

Table 2.2 shows that in 1993 there were marked variations in the levels of GIS adoption and automated mapping facilities according to type of authority. The highest level of adoption was in the shire counties and Scottish regions, whereas the biggest increase in take-up had occurred within shire districts. Further analysis revealed an increasing emphasis on departmental applications in both shire counties and Scottish regions, and on multidepartmental applications in the district authorities. In a third of the multidepartmental facilities, planning was the lead department. Planning

Table 2.1   Plans for GIS in Local Authorities in Great Britain

Plans for GIS	1991 Number	1991 Percentage	1993 Number	1993 Percentage
Already have GIS facilities	85	16.5	149	29.0
Plans to acquire GIS within 1 year	44	8.6	50	9.7
Considering the acquisition of GIS facilities	227	44.2	139	27.0
No plans to introduce GIS	158	30.7	176	34.2
TOTALS	514	100.0	514	99.9

Source: Masser, I. and Campbell, H.J. (1994) *Association for Geographic Information Conference Proceedings,* 14.2.1–14.2.6, London: AGI.

Table 2.2   Adoption of GIS by Type of Local Authority — 1993

Type of Authority	Number with GIS	Percentage
Shire districts	61	18.3
Metropolitan districts	34	49.3
Shire counties	43	91.5
Scottish districts	5	9.4
Scottish regions	6	66.7
Scottish islands	0	0.00
TOTALS	149	29.0

Source: Masser, I. and Campbell, H.J. (1994) *Association for Geographic Information Conference Proceedings,* 14.2.1–14.2.6, London: AGI.

departments also accounted for a third of all single-department GIS facilities, with other major single users being highways and engineering, emergency services, combined technical services, and estates departments. Compared with 1991, planning departments had strengthened their positions both as lead and single-use departments. Apart from emergency services, which increased in shire counties, the shares of all other departments either remained static or fell between 1991 and 1993, the highways departments being the biggest losers (Masser and Campbell, 1994).

More details of local authority GIS usage in England can be obtained from the two surveys undertaken by the Construction Industry Computing Association in 1992 and 1994 and summarized in Erik Winterhorn's paper to the 1994 AGI conference. These postal surveys produced a limited response but were supplemented by a survey of vendors early in 1993. The most common uses of GIS reported by English counties were traffic accident analysis, highway maintenance management, and map storage and production (Winterkorn, 1994). By far the heaviest uses of GIS by shire district councils related to map production, planning activities (particularly constraint mapping), and maintaining land and property records. The surveys found that GIS was most commonly used by metropolitan authorities for OS map production and maintenance and for processing planning application data. Box 2.3 illustrates that most of the local government uses of GIS in the early 1990s were for activities that were previously handled manually. Nevertheless, there were perceived benefits that included improved data storage, access, manipulation, and display; better access to and the integration of multiple datasets; and operational benefits due to improved

Box 2.3   GIS Applications in English Counties and Districts, 1992–94

English County Councils

The most common uses were:
- Traffic accident analysis, highway maintenance management, map storage and reproduction

Other applications mentioned included:
- Transportation and traffic flow analysis, highways and road inventories, planning/strategic auditing, development progress systems, relating land parcel roads to development, planning constraints, census data analysis, environmental auditing, crime recording and analysis, emergency management, county property databases, property management, footpath and public rights of way records, analysing and clients (including library catchment area analysis), mapping archaeological finds, recording commons and greens, dealing with mineral site applications, and keeping street lighting and streetworks records

English Shire District Councils

The most common uses were:
- Map production, planning activities (particularly constraint mapping), and maintaining land and property records

Other uses included:
- Zoning, calculating road lengths and parking areas for consultation and contract purposes, logging litter and street cleansing complaints (for monitoring contractors' performance, identifying and recording residual housing land, recording conservation area boundaries, maintaining sewer records, grounds maintenance, raising histories for planning applications, land charges searching, logging potentially contaminated land, recording listed buildings and ancient monuments, handling TPOs, and keeping records on housing stock and other buildings

Metropolitan District and London Borough Councils

The main uses were:
- OS map maintenance and map production and for processing planning application data

Other uses included:
- Storing land surveys, monitoring land use and property changes, property management, dealing with property enquiries, land terriers, processing land charges data, street gazetteers, highways maintenance, parks and grounds maintenance, registering contaminated and vacant land, recording land use constraints, and analysing census data

Source: From Winterkorn, E. (1994) *Association for Geographic Information 1994 Conference Proceedings,* 14.3.1–14.3.3, London: AGI.

performance in terms of speed of decision making, better management, and greater efficiency. The problems identified were complex and included sophisticated software, which required training to use, networking problems, and the costs associated with both purchasing digital data and customizing the systems (Winterkorn, 1994).

The signing of the SLA between OS and local authorities in 1993 was a major boost to the development of GIS, so much so that David Rhind, then the OS director-general, reported to the AGI conference in November 1995 that 80% of local authorities were using OS digital data compared with only 20% in 1993 (Rhind, 1995).

During the first half of the 1990s, other events triggered the development of GIM in local government. These included the changing needs and expectations of citizens, the growing concern about the environment — prompted by the Rio Summit in 1992 and emergencies like the Chernobyl incident in 1986 — as well as new legislation. The Computerized Street Works Register formally commenced on September 1,

1994, and encouraged both the development of GIS in many highway departments and closer work with the utilities.

By the time the AGI survey was undertaken in 1996, most U.K. local authorities were able to record the development of at least one GIS within their organization (AGI, 1996), and local government was being described as a major user of GIS. However, the reality was different, as GIS use highly concentrated within a very small population. The results of the AGI survey, in which the average authority had fewer than eight seats, were matched by the Royal Town Planning Institute's (RTPI) "state of the nation" survey in the autumn of 1995. This showed that over 70% of local-authority GIS had five seats or fewer, with 26% having only one. The RTPI survey also established that despite high take-up figures — 91% in counties — GIS was clearly still in its infancy, with only 30% of those authorities with GIS stating that they were fully operational (RTPI, 1998).

In 1998, the AGI director, Shaun Leslie, speculated on the reasons for the shallow GIS penetration in local authorities and concluded that:

- The benefits of GIS tend to be medium-term rather than short-term — unattractive to authorities driven by the need to balance the books during the financial year.
- Local government remained traumatized by a decade of continuous capital constraints which meant GIS funds were very limited.
- The GIS community had failed to capture the imagination of politicians, and had promised much but delivered little.
- The major block to take-up was the cost of data capture (Leslie, 1998)

2.6 WHAT PROGRESS WAS MADE TEN YEARS AFTER THE CHORLEY REPORT?

Recognizing that 10 years is a long time in a field as fast moving as geographic information, the AGI organized a symposium, "The Future for Geographic Information: Ten Years after Chorley," at the Royal Society on May 1, 1997. Its aims were to identify the progress made, establish what impediments to progress still existed, and look to the future. The issues debated are ably distilled by Ian Heywood in the AGI publication *Beyond Chorley: Current Geographic Information Issues* (Heywood, 1997), and some of the key points of particular relevance to local government are summarized below.

Ten years on, the OS had completed the coverage of the U.K. with vector digital map data and changed from a map producer and publisher to a provider of electronic data with a suite of high-quality products. By 1997 local authorities had benefited from 4 years' experience of operating with the SLA. However, global positioning systems (GPS), satellite imagery, and laser range-finders were starting to be attractive alternatives for capturing digital topographical data.

By 1997, Bill Gates's vision of "a personal computer on every desk and in every home" was coming to fruition. Over 90% of all PCs and workstations were using the Microsoft Windows interface, and GIS users had become familiar with how it looked and functioned (Wild, 1997). The impact of the Internet on GIS had been both sudden and dramatic. It started to break down the typical barriers for distributing spatial data, with networking growing in importance. The three major trends of connectivity, integration, and popularization led Vanessa Lawrence (1997) to describe the Chorley symposium as the dawn of the second age of GIS.

In the decade following the Chorley report, there was a dramatic growth in the number of higher educational institutions offering GI education either as part of their degree courses or as postgraduate programs. Over the course of the decade, there had also been some attempt to include GIS in the national school curriculum, although tight competition for funds in schools meant that the practical exposure of pupils to GI technology was limited (Heywood, 1997). The AGI had also been active in moves toward professional development for the GI industry. So, while education had responded moderately well to the Chorley report's original challenges, a long agenda still remained, including the need to increase awareness of the benefits of geographic information among managers and high-level decision makers, as well as among students in schools and universities.

2.7 HOW HAS THE MODERNIZING GOVERNMENT AGENDA AFFECTED GIS DEVELOPMENT?

On the same day that the Chorley symposium was held, a new government was elected with an agenda to modernize government and to use information and communications technology to meet the needs of both citizens and business. This brought new challenges and opportunities for the development of GIM in local government.

The modernizing government program set both the vision and the target for the capability for all government dealings to be electronic by 2008. In March 2000, the British prime minister, who championed the implementation of information age government at all levels, launched a new drive to speed up the process and brought the target date forward to 2005. In September 2000, he launched a £1 billion drive to get the U.K. online with the aim of ensuring universal access to the Internet by 2005, getting all government services online, and making the U.K. one of the world's leading knowledge economies. There are increasing indications that government at all levels has begun to recognize the importance of GI. Central and local government programs involving GI that target the high-priority issues facing society include neighborhood renewal, crime and disorder, agriculture, transportation, health, and quality of life.

Other aspects of the modernizing government approach, for example, the devolution of power to Scotland, Wales, and the London Assembly, the moves toward other regional assemblies in England, and changes to decision making in local authorities, all require underpinning by GIS with the ability to share information across organizational boundaries. Spatial visualization and Web technology will be particularly important in achieving joined-up geography and decision making in urban renaissance and community strategies.

The Improvement and Development Agency for Local Government (IDeA), the body that succeeded the LGMB in 1999, also recognizes that sound IM processes are essential steps toward the best value. Following a proposal in the *Modernising*

Government white paper (Cabinet Office, 1999), the IDeA jointly formulated a Central/Local Information Age Government Concordat to encourage innovation and cooperation. This was agreed with the government and the Local Government Association in July 1999.

All this change has helped to generate significant growth in local government GIS users and, with the help of Web technology, many authorities have developed public information systems. In 2000, 56% of local authorities had a fully operational GIS with a further 38% in development. An increase had also occurred in the number of individual licensed GIS users with half the authorities having between 11 and 50 seats, and some with more than 50 seats (RTPI, 2000).

As the new century began, GIS was entering the mainstream as part of a suite of IM systems, characterized by hugely improved interfaces, interchangeable data, and global networking. GIS had ceased to be the domain of the specialist and had become a more universally available and applied tool with the accent firmly on the user and the customer (Gill, 2000).

2.8 WHAT LESSONS CAN BE LEARNED FROM THIS REVIEW OF GIS DEVELOPMENT?

The main lessons that can be distilled are that:

- Most early applications focused on automating those operations that traditionally had been performed manually and were frequently perceived as expensive, inflexible, and difficult to use.
- The search for a single spatial referencing system for each of the the U.K.'s properties began in the early 1970s, but did not really take off until the mid-1990s.
- The publication of the Chorley Committee report in April 1987 raised the whole profile of IM in local government.
- The time and resources required to capture data and the failure to capture the imagination of sufficient senior managers and politicians have often been significant barriers to the development of GIS.
- Human attitudes, perceptions, and knowledge have not kept pace with the rapid technological progress, with the result that organizational factors are more likely to affect the success or failure of a GIS than technical issues.
- The signing of the first SLA between the OS and local authorities in 1993 was a major boost to the development of GIM in local government.
- Government leadership and the various coordinating mechanisms have vital roles to play in the development of GIM; particularly important are the twin drivers of best value and joined-up electronic modernized government.

PART 2

Key Elements of Geographic Information Management

CHAPTER 3

Organizational Content

KEY QUESTIONS AND ISSUES

- How important is the organization to the success of GIS?
- What are the government's expectations of local authorities?
- What are the main drivers for change in local government?
- Who are the main users of GIM in local authorities, and what are their needs?
- What are the arguments for a corporate approach?
- Where has GIS been used to greatest effect in local government?
- What has constrained GIS potential in local government?
- What lessons can be learned both from local government and other organizations?
- What organizational changes are likely to result from GIS development?

3.1 HOW IMPORTANT IS THE ORGANIZATION TO THE SUCCESS OF GIS?

Most definitions of GIS focus on the hardware, software, data, and analytical processes. However, no GIS exists in isolation from its organizational context, and this is a particular aspect of emphasis within this book. There must always be people to plan, implement, and operate the system as well as make decisions based on the output (Heywood, Cornelius, and Carver, 1998). While all U.K. local authorities operate within the same basic legislative framework, each authority is unique with its own agenda and its own way of doing things based on its traditions, culture, style, responsibilities, and external pressures. This emphasizes the importance of looking at the issues from the perspective of how organizations actually operate rather than a hypothetical notion of how they should (Campbell and Masser, 1995). Speaking directly to the reader who is involved in implementing GIS within local government, the essence of a successful GIS is to start by thinking about your own local authorities and their citizens, about their information needs, and how many of these have a spatial dimension. In John England's words, "Do not think of system, think information" (England, 1995).

As the limitations of technology recede and geospatial digital data become more widely available, the impact of organizational factors on the success or failure of GIS achieve greater prominence. In fact, Derek Reeve and James Petch (1999) conclude that:

> Building a successful GIS project depends at least as much upon issues such as marshalling political support within the host organization, clarifying the business objectives which the GIS is expected to achieve, securing project funding, and enlisting the co-operation of end-users, as upon technical issues relating to software, hardware, and networking.

Organizations exist because one person cannot do everything. They develop their own cultures and structures, they develop their own ways of doing things, and they contain both formal and informal groupings, often with their own aspirations. The "what's in it for me" factor can have a powerful influence on the implementation of GIS, especially when the authority's GIM objectives are not specifically stated.

We believe that clarifying the organization's needs and aspirations are vital to the success of any GIS and must be addressed at the outset. That is why we begin this review of the components of GIM by examining the organizational context. Our thought processes have been influenced by the academic writings of Bob Barr, Michael Batty, Heather Campbell, Ian Heywood, Ian Masser, Derek Petch, James Reeve, Michael Worboys, and others, as well as our practical experiences from within local government. But we start by looking at the expectations of central government.

3.2 WHAT ARE THE GOVERNMENT'S EXPECTATIONS OF LOCAL AUTHORITIES?

The "e-government strategy" of the U.K. contains an ambitious agenda to achieve the prime minister's vision of modernized, efficient government alive to the latest developments in technology and to meeting the needs of both citizens and businesses — an agenda that will not be achieved without considerable effort, investment, and cultural change. It fulfils the commitment in the *Modernising Government* white paper to publish a strategy for information age government and has four guiding principles (Cabinet Office, 2000):

- Building services around citizens' choices (people should be able to interact with government on their own terms)
- Making government and its services more accessible (all services that can be electronically delivered should be)
- Social inclusion (including a commitment to make it easier for all people to access the Internet, whether individually or through community facilities)
- Using information better (recognizing that the government's knowledge and information are valuable resources)

The government recognizes that implementing the strategy will place significant demands on all public servants to work in new ways and to acquire knowledge about the new technology, and will have implications for policy making, service delivery, management, and organizational culture.

The publication of the *Modern Local Government — In Touch with the People* and *Local Voices: Modernising Local Government in Wales* white papers in 1998 set in train a number of changes to modernize councils in England and Wales. Of particular relevance to the management of GI are:

- Improved local services through best value
- New models of political management clearly separating the councils' executive and representative roles
- A new power of well-being that allows councils to find innovative ways to meet their areas' social, economic, and environmental needs
- The ability to prepare community strategies (DETR, 1998)

A new duty of best value was introduced by the Local Government Act 1999, requiring all councils to fully appreciate the clients' needs and involve them in the democratic processes governing service delivery. This was followed by the Local Government Act 2000, requiring all councils to adopt new arrangements for making decisions, giving them new powers to promote community well-being, and enabling them to work in partnership with the business and voluntary sectors to develop visions for their communities. Underpinning each of these tasks is the need to assemble and share vast quantities of GI.

The Information Age Government: Targets for Local Government white paper recognized that local authorities were at very different levels of development in their approaches to electronic service delivery. Therefore, they encouraged authorities to set their own targets in relation to best value performance indicators, stressing the importance of joined-up seamless services (DETR, 2000). The white paper emphasized that nationally coordinated projects like the National Land and Property Gazetter and the National Land Information Service, were needed to develop the underlying infrastructure necessary for councils to provide joined-up service delivery.

The U.K. government's vision for the e-citizen is the ability to access online government services 24 hours a day, 7 days a week. The prime minister has issued a challenging target proposing that all dealings of the public with local government should be capable of electronic delivery by 2005. The e-government vision requires a massive change of culture on the part of local government, though doubts have been expressed about whether the 2005 target is realistic and practical without substantial extra funding (Adnitt, 2000).

3.3 WHAT ARE THE MAIN DRIVERS FOR CHANGE IN LOCAL GOVERNMENT?

In addition to the government's expectations, there are a number of drivers for changing GIM in local government. These arise from both internal and external sources.

The internal drivers include:

- The increasingly commercial and business-like approach of local authorities, including the emergence of more effective and coordinated "top-down executive leadership" with clear mission statements, corporate plans, and service delivery plans
- Better integration of corporate information avoiding data duplication and fragmentation as well as exploiting the data riches buried in the coffers of local authorities
- The desire to achieve efficiency savings by reducing the costs of data collection and the costs of maintaining up-to-date records
- The enthusiasm and commitment of champions and change agents — both staff and, less frequently, politicians
- More effective (and comprehensive) presentation of information to decision makers, including the improved ability to simulate the impact of policy choices, to decide priorities, and to monitor outcomes
- The desire to make GI more accessible to citizens

The external drivers include:

- The rapid growth of the Internet and the increased availability of digital data
- Low-cost hardware and user-friendly software with much improved performance
- The emergence of standards for data capture, manipulation, and exchange, including the National Transfer Format (NTF) now published as BS7567, and BS7666
- More demanding electorate, growing public expectations, and increasing spatial awareness
- New momentum in the regional and environmental agendas
- Increasing need for collaboration and information sharing between local government and other organizations on spatial issues, such as crime and disorder, street works, the provision of welfare services, and the handling of emergencies

As with all new technology, it often takes one or two champions in the organization to get GIS going. These champions can be politicians as well as officers and are usually people who combine a knowledge of the organization and its processes with an innovative character, a keen interest in modern information technology, and an urge to move forward. In contrast to the more conservative members of the authority, they generally form a small minority. However, the key to their success lies in spreading their enthusiasm and commitment to other colleagues.

3.4 WHO ARE THE MAIN USERS OF GIM IN LOCAL AUTHORITIES, AND WHAT ARE THEIR NEEDS?

GIM embodies a vast array of perceptions and a wide range of users each with their own needs. Several writers have used different user classifications, but we find Jan Roodzand's most helpful in a local government context. He divides users into three broad groups: *viewers:* those who view the information on an *ad hoc* basis; *users:* those who need access to the data for day-to-day activities; and *doers:* those who have strong skills in GIS and data management (Roodzand, 2000). Similar classifications are used by Intergraph (Wild, 1997, and Hoogenraad, 2000) and others.

The first group, the viewers, is by far the largest, and they use the results of geographic analysis for a variety of often unspecified reasons. Most of the time this group views the information only to get the answers to fairly basic questions. In general, their insight into spatial data and GIS is minimal, so retrieving the required "picture" or analysis of the area they want to view should be as easy to access as the latest plug-and-play PC technology. According to Lawrence and Parsons (1997), this group's members frequently use "GIS in disguise." In other words they are not even aware they are using GIS. In recent years, Internet and intranet technologies have played increasingly important roles in enabling the viewers to access the data they require via simple application interfaces. In local government, this group comprises a diverse range of people from councilors, senior and middle managers, professionals and technicians, administrative and clerical staff to citizens and community groups.

The second group represents those professional and administrative users who benefit from access to spatial data in their daily activities. Although they may work with GIS up to 20% of their time, their technical knowledge of the systems' complexities is still limited. However, this group is familiar with spatial data and knows how to interpret it. They need to analyze data and can be expected to handle complex queries. They require immediate access to GI and easy-to-use software that are integrated with their office software. They are often served with the more comprehensive Internet and intranet or client-server applications.

This group is comprised of many local government's traditional map users — planners, engineers, surveyors, valuers, and so forth — together with a growing number of staff and councilors who have more recently recognized the advantages of GIM. However, a 1999 survey of professional planners has shown that while virtually all members of this group had access to a computer at work for word processing, only 11% used computerized GIS on a daily basis (RTPI, 1999). As planners are frequently those who take the lead in promoting GIS development in their authorities, there is no reason for assuming that the percentage will be any higher among other professionals.

The viewers and users need the much smaller but vital group of GIS experts to enable them easy access to the information they need. These experts are the data managers — the doers — who are responsible for ensuring that the geographic information is readily available and of the right quality. These are highly skilled individuals who understand the complexities of GIS technology, data accuracy and consistency, design and maintenance rules, and approaches to implementation. The doers tend to spend over 80% of their time creating, maintaining, and managing spatial data. To accomplish this, they require a clear view of the authority's corporate GI needs and the resources to acquire and maintain both the spatial data as well as the computer systems to serve the specialist needs of all parts of the organization.

The doers are limited in number but are essential to the development of GIS in any authority, whether they are employed by that authority or provided by a consultancy.

3.5 WHAT ARE THE ARGUMENTS FOR A CORPORATE APPROACH?

One of the most frequently asked questions about GIM in local government concerns the relative merits of the departmental and the corporate approaches.

Looked at rationally, the corporate approach has several advantages including maximizing returns on the investment and making common databases available to the largest possible number of users, thereby avoiding duplication and resource waste. While departmental solutions are initially cost effective, as they multiply, so can the cost of duplication of activities, training, data, and maintenance.

Derek Reeve and James Petch (1999) argue that such corporate rhetoric may be attractive, but often organizations comprise loosely related "fiefdoms" engendering "turf protection" rather than rational cooperation. U.K. local authorities carry out a disparate range of activities allocated to them by central government. Many of these are separate operations with little in common, each requiring different professional skills, operating under different legislation, and having different budgetary requirements. According to Reeve and Petch, "To expect such disparate activities to dovetail easily into a coherent corporate whole is often unrealistic."

The point that these and other commentators make is that in the world of real organizations the need for a corporate approach can be pushed too far. Experience shows that although there were some brave attempts at setting up corporate GIS in the early 1990s, based on the limited rollout of expensive systems, few systems were truly corporate. This can occur only when GIS penetrate every workplace and top managers take ownership (Gill, 1998). Most success has been achieved through information sharing both within authorities and with external bodies.

However, as per Gill (1998), corporate GIS do encourage:

- Data to become a corporate issue
- Data sharing to reduce wasteful duplication
- An open and collaborative attitude toward sharing spatial information
- Corporate standards to ensure maximum benefits
- Increased lateral communications and team- or project-based working

On the other hand, the advantages of the departmental approach, involving separate systems development, are increased independence and sensitivity to user needs, clearer lines of responsibility, and closer control over priorities. However, while the departmental approach can sometimes achieve quicker initial progress, it can often run into loss of momentum as conflicts with incompatible GIS initiatives in other departments arise.

The corporate approach to managing information is an e-government target, and technology can now enable councils to link their systems so that they can better support joined-up services. The approach that each authority adopts depends on its own particular circumstances and style. The case studies in Part 3 contain both corporate and departmental approaches and draw out the lessons learned from both options.

3.6 WHERE HAVE GIS BEEN USED TO GREATEST EFFECT IN LOCAL GOVERNMENT?

GIS have roles to play at all levels in local government, from basic map handling through to complex spatial analysis and decision support. There is a growing recognition that GIS are not simply systems for automating manual tasks or speeding up data handling, but are "integrating frameworks" that can make an enormous contribution to improving both the efficiency and the effectiveness of an authority.

Given that GIS users in local government historically have been concentrated in the planning, engineering, and estates services, it is hardly surprising that GIS have been used to greatest effect in the following activities:

- Flexible mapping
- Land and property matters
- Network analysis
- Incident analysis
- Socioeconomic analysis
- Environmental monitoring and management

Box 3.1 illustrates that within these broad, and sometimes overlapping, headings there are a wide range of uses and applications, several extending beyond the remit of the services listed above. Many of these, like map production and maintaining land and property records, focus on achieving efficiency savings where the main

Box 3.1   Where GIS Are Used to the Greatest Effect in Local Government

Flexible Mapping
- Consistent and easy-to-find maps for all services
- Automated updating of base maps
- Easier map production and plan processing
- Seamless customized maps
- Base maps plus user overlays

Land and Property
- Land and property gazetteers
- Planning applications and local land charges
- Planning constraint and policy areas; land use and terrain analysis
- Asset management, including grounds and roadside maintenance
- Identifying unused, underused, derelict, and contaminated land
- Locating sites for housing, schools, mineral extraction, waste disposal, etc.

Network Analysis
- Roads management (e.g., inventories, assessment of condition, maintenance)
- Accessibility and route planning (e.g., waste collection routes, transporting clients to day centers and children to school, supply delivery routes)
- Coordination of street works
- Pipelines and power lines

Incident Analysis
- Traffic accidents, holes in the road, street lighting faults
- Drugs, crime, and disorder
- Environmental health, noise, litter, and other complaints
- Pollution incidents, health epidemics, and other emergencies

Socio-Economic Analysis
- Population analysis (structure, numbers, location, characteristics, etc.)
- Citizen profiling (geodemographics)
- Facility planning and catchment area analysis
- Assessment of housing and leisure needs

Environmental Monitoring and Management
- State-of-the-environment reports and Local Agenda 21
- Archeology, landscape, and ecology
- Listed buildings, sites of special scientific interest, and conservation areas
- Impact assessments (e.g., wind farms and large structures)

Figure 3.1   GIS benefit profile.

objective is to reduce operating costs as computer systems replace manual methods. Others lead to much improved effectiveness as GIS extend operational capabilities, facilitate the integration of services, and improve the quality of decisions made. These benefits are illustrated in the profile set out in Figure 3.1.

Increasingly, senior managers and politicians in local government are looking for GIS applications where there is a clear citizen focus and a clear business case. The growing number of computerized local land charges systems is a good example. Here, the time taken to handle requests has been reduced from days to hours with benefits both to those involved in property conveyancing and to the efficiency and effectiveness of the council.

Figure 3.1 indicates that senior managers and politicians have a particular interest in achieving best value, improving the authority's image and services to the public, and using corporate information to make their executive and strategic decisions. Recent developments in Web-based technologies — both Internet and intranet — have done much to improve the effective use of GIS in these areas. A report published on the World Wide Web by UKFavourites.com (2000) contains various examples ranging from reporting street lighting faults over the Internet in Knowsley through online access to committee reports and structure plan policies in Hampshire and Devon, respectively, to the opportunity to join public discussion forums in Cumbria. Of particular note are the interactive planning register in Wandsworth and the Local Agenda 21 network in Lancashire; more details about these are given in Box 3.2.

However, the report concludes that "whilst most local authorities have Websites, the levels of interaction and participatory tools are extremely limited at present" (UKFavourites.com, 2000). Thus, there is still a long way to go before the government's electronic service delivery targets are realized.

3.7 WHAT HAS CONSTRAINED GIS POTENTIAL IN LOCAL GOVERNMENT?

Over the last 10 years, a number of surveys and research projects have explored the constraints that have limited the potential development of GIS in local

Box 3.2   Interactive GIS Networks in Wandsworth and Lancashire

The best example of a local council offering an all round electronic planning service has to be the *London Borough of Wandsworth*. Not only did this Council create the first interactive planning register, copied by many authorities, it is continuing to develop its service admirably. The amount of planning information is astounding, their adopted and draft revised Unitary Development Plan is online and comments can be made electronically, users can query the latest planning news and search for details on any planning application going back to 1947. The public can submit a planning application online or make comments online. The next stage to create a truly interactive electronic service would be for Wandsworth to allow payment for planning applications fees online via encrypted credit card payments.

Local Agenda 21 has been a test bed for new ways of engaging the public through new initiatives such as electronic forums. These forums give the public degrees of citizen power through partnerships with the policy makers. *Lancashire County Council* are one authority commended for its electronic forum on its Local Agenda 21 Internet Website (http://www.la21net.com). Its innovative interactive facilities include discussion groups, a comment book, an interactive database of local companies, and online complaints. Lancashire County Council had received over 100 emails from businesses in Lancashire praising their interactive company database. This allows the user to tailor the information to their needs rather than wade through lots of irrelevant facts.

Source: http://www.ukfavourites.com.

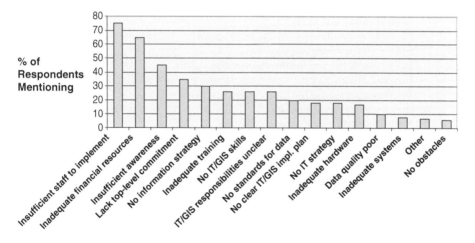

Figure 3.2   Obstacles to effective IT and GIS. (*Source:* Based on RTPI, 2000.)

government and examined why GIS diffusion has not been as great as first antici-pated. These results have been combined with our practical experience of GIM in a range of local authorities to produce the following list of constraints. To indicate the order of importance of the main obstacles, we have compiled Figure 3.2, which summarizes the results of the RTPI survey of IT in Local Planning Authorities in 2000 (RTPI, 2000).

The various restraints limiting GIS development are:

Inadequate Resources
Insufficient staff numbers to implement and operate the GIS
Inadequate financial resources resulting from both capital and revenue constraints, especially in smaller authorities

Insufficient Awareness, Poor Perceptions, and Lack of Commitment
Lack of top-level commitment
General lack of IT awareness and of GIS potential in particular
A bad GIS experience from an earlier involvement
Failure to capture the imagination of politicians or lack of credibility of early GIS decision support systems or both
Skeptical or conservative perceptions of other stakeholders such as senior managers, staff, and citizens
Organizational instability resulting from changes in political control, local government reorganization, or both

Lack of Strategy
No information strategy, leading to lack of vision, imagination, and innovation
Lack of a business case
Neglect of the human issues — too much technical emphasis
Lack of IT strategy resulting in an inadequate technical context within which GIS can be implemented and operated
No clear plan for implementing GIS
GIS not integrated into the decision-making culture

Inadequate IT and GIS Skills and Support
Responsibilities for IT and GIS support unclear
Lack of IT and GIS skills
Champions and pioneers moving on to another organization
Enthusiasts providing answers to questions nobody may ask
Departmental barriers to effective communication
Personal decisions of staff that delay, hinder, and frustrate rather than help
Insufficient or inadequate user training

Technology and Data Problems
Inadequate hardware
Key systems do not meet requirements
No standards for data
Digital data, although available, are not affordable
Not perceiving data as a corporate resource

3.8 WHAT LESSONS CAN BE LEARNED BOTH FROM LOCAL GOVERNMENT AND OTHER ORGANIZATIONS?

Experiences within local government, a literature review, and direct contact with a wide range of people involved in GIM have led us to conclude that many important lessons can be learned from the progress already made by a wide range of organizations. We share these lessons with the reader at this stage of the book and will return to them as we examine the other components of GIS and analyze the case studies in later chapters.

First of all, our overwhelming conclusion is that no single model or blueprint can be applied to every organization. We agree with Gill (1998) that just as each local authority has developed its own strategies, structures, systems, and style, so there will be many different approaches to the development of GIS. In fact, Campbell and Masser (1995) advise us to beware the rational "cookbook" approach and suggest that there are no quick fixes or recipes for success. They conclude that there are three important areas that invariably require careful consideration: the information management strategy, commitment of individuals at all levels, and an ability to cope with change. We list each of these to highlight the lessons learned so far.

Information Management Strategy

Develop a geographic information strategy that brings together corporate initiatives and government ideals, clarifies business objectives, and sets out what GIS is expected to achieve.

Ensure that an IT strategy exists to provide a clear technical framework for the implementation of GIS (or if not, that at least any mandatory IT standards and constraints are clearly identified).

Enlist the cooperation and involvement of end users and identify their information needs.

Plan for the big picture and think corporately even if the early applications are departmental.

Be brutally realistic about the short-term costs while recognizing that many of the benefits will be realized in the future.

Set clear time-scales for producing the results, including short-term milestones for early deliverables.

Keep data conversion at the heart of project planning and management.

Commitment of Individuals at All Levels

Marshal political or senior management support to obtain a clear mandate and to secure sufficient project funding.

Talk business, not technology, and be realistic.

Identify an effective project champion at a senior level and motivate the other stakeholders.

Appoint a GIS manager (the most important member of the GIS team).

Ensure that the end users are on board; aim to win over the skeptics.

Establish effective partnerships between users and IT and GIS specialists, rather than just contractual relationships.

Coping with Change

Identify and plan the key stages of the project — be realistic about time and performance expectations.

Establish a process for managing change and transition, including effective project management.

Document a change management process and get all parties, including the unions, to agree with it before project implementation begins.

Form a management steering committee and put management, not project staff, in control of the change.

Arrange user briefings, demonstrations, and training sessions.

Keep changes to work practices and job descriptions under review.

3.9 WHAT ORGANIZATIONAL CHANGES ARE LIKELY TO RESULT FROM GIS DEVELOPMENT?

It isn't the strongest or even the most intelligent that survive but the ones that adapt the most quickly to change.

—*Charles Darwin*

GIS involve such different and innovative ways of working that there are bound to be many cultural changes, not just at the individual human level but also at the organizational level. (Campbell and Masser, 1993 and 1995, quoted in Gill, 1998).

While change and uncertainty are integral parts of organizational life, any GIS implementation will impact work practices, processes, information flows, management structure, staff, and organizational culture. The most significant aspect of introducing a new technology is the extent to which it implies change to existing practices, particularly the underlying approach to decision making. It is important to recognize that proposing corporate implementation of GIS may imply a major structural change to the way activities are conducted (Campbell and Masser, 1995). Because of varying values and practices, there is not a single approach to implementation that guarantees effective utilization in all circumstances. Even departmental GIS can have far-reaching effects on an authority's structure and culture.

Just as IT facilitates the spatial disaggregation of activities that previously had to be located together for purposes of control and coordination (e.g., data processing and word processing in the Far East, more people working from home), GIM could also affect where and how people operate in local government. Reeve and Petch (1999) argue that once a local authority has all of its planning applications data on GIS, there is no need for development control officers to continue working from central council offices. They contemplate that many planners will soon be able to take their departmental databases on-site using hand-held devices, and this could clearly apply to other professionals. Indeed, many surveying and utility companies are already using mobile GIS and GPS systems in order to maximize the time their field operatives spend away from their office. In recent years, an increasing number of local authorities are using on-site systems. Cheshire County Council is one example that has used a field solution from Positioning Resources to help maintain the county's rights of way records.

Any GIS project will impact the shape and culture of the organization, and a number of learned papers have recommended that GIS implementers adopt the business process reengineering (BPR) approach. Aybet (1996) argued that the low success rate achieved by local government GIS projects occurred because they were not implemented in the correct manner and advised that greater success might be attained if their implementation reflected a business reengineering perspective. However, not only has BPR been frequently associated with downsizing, it has sometimes treated people as if they were "just so many bits and bytes — interchangeable parts to be reengineered" (Reeve and Petch, 1999). For GIS to be successful, the staff involved must be committed to making it so. If the staff feel threatened by the speed and severity of the changes they are experiencing, they may be obliged to accept them, but compliance is not the same as commitment. Therefore, rather than adopting a radical BPR-style approach to GIS implementation, it is better adopt an evolutionary style allowing the users to realize the benefits at an acceptable pace.

As GIS become a corporate service, it is important to decide where the GIS unit should reside (Campbell and Masser, 1995). Traditionally, there have been three options: in a computer services department, at the executive level, or in one of the user departments. Whichever option is chosen will result in organizational changes and affect information flows.

Information flows and management structure are closely related. Reeve and Petch (1999) argue that the relevance of the familiar triangular organizational model is being rapidly undermined by the widespread introduction of information systems. They contend that these systems tend to blur rigidly hierarchical reporting structures. IT senior managers can tap directly into corporate databases, executives can dip into electronic mail at any level, and relatively junior officers can send messages to everyone on the system. As a result, middle-ranking officers lose power to higher-level officers and politicians, leading to "flatter" organizational structures. The electronic government vision and the rapid development of Web technology are tending to reinforce this trend.

There is little doubt that the continuing development of GIM will give rise to a change in culture and the way a local authority operates on a daily basis. This is a long-term process and requires leadership from the highest level combined with understanding and commitment at all levels of staff. Managing these changes through GIS implementation will be picked up again in Chapter 6. In the meantime, we turn our attention to considering the three main elements or legs of GIS.

CHAPTER 4

Spatial Data

KEY QUESTIONS AND ISSUES

- What are the main characteristics of spatial data?
- What are the main types and sources of spatial data?
- What is a data model and how is spatial data modeled?
- What methods of data capture are available?
- What types of databases are used in GIM and why are they so important?
- Why is data quality important and how do we achieve it?
- What analyses are typically carried out on spatial data?
- How do models of spatial processes help decision making?
- What are the main forms of GIS output?

4.1 WHAT ARE THE MAIN CHARACTERISTICS OF SPATIAL DATA?

GIS are simplified computer representations of reality. The data they use are typically observations and measurements made from monitoring and recording the world around us. However, capturing the appropriate data can be a daunting and time-consuming task. Although there are many sources, there are basically only two categories: primary data, collected through first-hand observation, and secondary data, collected by another individual or organization.

All data typically have three dimensions relating to their location (where they are), their attributes (what they are), and the date when they were collected. GIM places the greatest emphasis on using the locational or spatial element for transforming data into information, thereby giving it meaning. As we have seen already, the traditional way of storing, analyzing, and presenting spatial data is the map. Cartographic methods are centuries old, and there are many similarities between their approach and the theoretical framework for GIS. Hence there is a great deal to learn from the cartographer's approach, not least that the purpose of the map decides the features to select and defines the amount of generalization, the spatial referencing system, and the method of representing of the data.

During the mapping process the cartographer must:

- Establish the purpose the map is to serve
- Define the scale at which the map is to be produced
- Select the features (spatial entities) from the real world that must be portrayed on the map
- Choose a method for the representation of these features
- Generalize these features for representation in two dimensions
- Adopt a map projection for placing these features onto a flat piece of paper
- Apply a spatial referencing system to locate these features relative to one another
- Annotate the map with keys, legends, and text to facilitate use of the map (Heywood et al., 1998, after Robinson et al., 1995).

The scale of the map is determined by the purpose or purposes to be served and represents the ratio of a distance on the map to the corresponding distance on the ground. That is, at a scale of 1:2500, a line of 1 cm on the map represents a line of 2500 cm or 25 m on the ground. Local authorities use a wide range of map scales, but the most common are 1:1250, 12,500, and 1:10,000 for large-scale mapping and 1:50,000 for small-scale mapping.

Fundamentally, maps use three basic symbol types to represent real-world features: points, lines, and areas. The same three basic spatial entities are used in any GIS. Points are used to represent features that are too small to be shown as areas, e.g., lamp posts, manhole covers, and street furniture on large-scale maps. Lines, which are simply an ordered set or string of points, are used for linear features such as roads, pipelines, administrative boundaries, and river networks. Networks are sometimes treated as a separate data type but are really just an extension of the line type. Finally, areas are represented by a closed set of lines and are used to define features such as buildings, fields, and administrative areas. Area entities are frequently referred to as polygons. As with line features, some of these polygons exist on the ground, e.g., buildings, and some are imaginary, e.g., census enumeration districts. Three-dimensional areas are treated as surfaces, which can be used to represent topography or nontopographic features such as pollution levels and population densities. Sometimes, surfaces as well as networks are considered as separate entity types.

Each spatial entity may have more than one attribute associated with it. Attributes are the nongraphical characteristics of the entity. For example, they can describe the type of building defined by a polygon — a house, a school, or an office — or the class of road represented by two parallel lines. These attributes allow certain GIS operations to be performed, e.g., "where are all the primary schools within a particular ward?" or "which is the shortest route from A to B?" However, in order to answer such questions, the geometric relationships between the spatial entities must be understood.

In GIM, topology is the term used to describe the geometric characteristics of spatial entities or objects. In relation to spatial data, topology comprises three elements: adjacency, containment, and connectivity. Objects can be described as adjacent when they share a common boundary, whereas containment describes one feature contained within another, e.g., a house within a garden. On the other hand, connectivity is the geometric property used to describe linkages among line features, e.g., roads connected to form a bus network (Heywood et al., 1998).

In order to carry out analyses of the basic spatial entities, it is necessary to treat the spherical Earth as a flat two-dimensional surface (a sheet of paper) by using a suitable map projection. This transformation is achieved by approximating the true shape of Earth, thereby introducing errors into the spatial data. These will vary depending upon the projection method chosen from the wide range available. Some will distort distances, others direction, while others will preserve shape but distort areas. Users need to know which map projections are being used, particularly if they wish to combine data from different sources. Otherwise, features that exist at the same location on the ground may appear to lie at different geographic positions when viewed on the map or computer screen. For mapping small areas of the globe, especially those like the U.K. that have only a small extent of latitude, the Transverse Mercator projection is often used. It has the advantage of maintaining scale, shape, area, and bearings over small areas and was chosen as the basis of the OS's National Grid system.

Spatial referencing is used to locate a feature on Earth's surface or on a map. Several methods of spatial referencing exist, all of which can be grouped into three categories: geographic coordinate systems (latitude and longitude), rectangular coordinate systems (e.g., the OS's National Grid system), and noncoordinate systems (e.g., the U.K. postcode system). Most spatial referencing systems have problems associated with them. Heywood et al. (1998) list three examples: spatial entities may be mobile — e.g., animals, cars, and people can be located only at a particular time; spatial entities may change — e.g., road improvements occur, policy areas are redefined; and the same object may be referenced in different ways — e.g., a building may be represented as both a point and a polygon on maps of different scales. Despite these problems, the ability to link, or "glue" together, disparate datasets using spatial referencing is vital to the management of geographic information, as the following section will show.

4.2 WHAT ARE THE MAIN TYPES AND SOURCES OF SPATIAL DATA?

Data about local authorities' areas and activities are produced continuously. Many of their everyday activities produce spatial data automatically, some of which is stored digitally in databases but much of which still remains in analogue form in files, ledgers, and photographs. In addition, local authorities use data from various central government departments as well as aerial photography, satellite imagery, and field surveys.

Not only are there now an abundance of spatial datasets available both to local authorities and their citizens, there are a wide variety of sources providing data that differ widely in content, currency, and role. Writing in the *AGI Source Book for GIS*, 1997, Hugh Buchanan usefully categorized this data into three varieties (see Box 4.1):

- Application data that gives information of importance for answering a particular question
- Parcel data that describes abstract units of area that the world is divided up into
- Topographic data that tells you about the physical surroundings

Buchanan goes on to explain that, for many purposes, some data of each sort is required:

Users often already have some application data, and wish to relate it to some other application data, together providing the facts that are of most direct interest. These facts have to be attached or glued to each other, or alternatively to the real world. This is done by using some parcel data that relates the spatial content of some application data to the spatial content of other application data (for example postcodes to census areas). Additionally, it is usually useful to relate these parcels to the real world in the form of some topographic data, so that the data can be vizualized or inspected.

Box 4.1   Data Varieties

Application Data (Interest)

The term application data covers many things, such as socio-economic, geological or property data. A user will often have their own data (such as customer records), and is often also interested in adding value to their own information by relating it to other sets of data.

One major source of data about population is the (decennial) census carried out by the Office for National Statistics in England and Wales, the General Register Office in Scotland and the Census Office for Northern Ireland. In addition to the factual bones of the census, much socio-economic flesh is added by surveys of population and behavior. For other application areas, the required data will be different, such as geological, hydrological and land use data.

Parcel Data (Glue)

Socio-economic application data is often spatially described using a street address, a postcode, an electoral ward or a census enumeration area, but very rarely by a National Grid (map) co-ordinate. Land-related information is very often described by a National Grid co-ordinate, but may be described by an administrative area, such as a county. There are a variety of data products that relate one set of parcels to another and individual parcel sets to the National Grid.

Topographic Data (Real World)

Topographic data corresponds to the traditional published map, but is now available in a variety of different forms. The first of these is the vector map, where the co-ordinates of each line, point and piece of text are included.

A common alternative to vector maps are raster maps. The raster consist of a fine grid of cells, each of which carries a colour value. By displaying the raster, the user can recreate the type of visual appearance that a paper map would have had.

In recent years, a third form of topographic data has become increasingly common. This consists of photography and satellite imagery. In computer readable form, these types of data are raster. They are created from cameras and other sensors carried by aircraft and satellites, and are very good at retaining the overall visual impression of the surface, since (for example) the nature of the ground cover can be seen on the image.

The largest supplier of topographic data in the U.K. is the Ordnance Survey, who have a wide range of data products. Other suppliers of such data are land survey firms who will create data to order, and other data publishers such as Bartholomews and the AA.

Source: Extracted from Buchanan, H. (1997) Spatial Data: A Guide, in D.R. Green and D. Rix (Eds.), *AGI Source Book for Geographic Information Systems 1997,* London: AGI.

In local government, the OS's digital topographic database provides the bedrock for GIS in the traditional map-using services like planning, highways, and estates. However, for many users aerial photographs are easier to interpret as they provide a real picture of the world at a known point in time. Raw photographs are not as accurate as maps as they contain scale distortions, especially at their edges, and make buildings appear to fall away from the center. This problem, together with errors due to changes in ground relief, can be resolved by a process known as orthorectification.

Increasingly available are off-the-shelf products containing aerial photographs that have been scanned, orthorectified, and stored as digital databases. The sources for this data include:

- Geoinformation Group, a U.K. company formed from a management buyout of Cities Revealed products, providing 25-cm digital databases corrected to OS mapping focusing on cities or counties in high-demand areas
- Getmapping.com (formerly Millennium Mapping Company) originally formed to create a millennium archive of the U.K. at 1/10,000 scale
- U.K. Perspective, a joint venture between NRSC and Simmons Aerofilms, providing another millennium archive with the ability to create digital orthophotographs on demand

For practical purposes, digital imagery is mainly used in a compressed format due to large storage requirements. For example, with the normal 25-cm resolution, a 1-km^2 tile takes approximately 45 MB of disk space. However, commercially available software such as Mr SID enable images to be reduced to about 2 MB without significant loss of clarity, making imagery considerably more manageable (Denniss, 2000).

High-resolution imagery is also available from satellites and new digital airborne imagers. This is invaluable not only in the construction of an accurate and comprehensive GIS database but also in maintaining the database at a reasonable cost. New sources of satellite information that are more affordable and have much improved ground resolutions are becoming available. Often the frequencies used to capture the data are such that they can penetrate cloud cover and the data can be quickly processed to order.

Land, property, and highways services often describe their data by National Grid coordinates, but most application data in local government is glued together by an address or the postcode system. As a result, local authorities have found both the OS's ADDRESS-POINT and the Royal Mail Postcode Address File (PAF) invaluable as a means of linking Great Britain's 25 million addresses and the unit postcodes to National Grid references. The Gridlink initiative launched at the GIS 2000 conference by the OS, the Office for National Statistics (ONS), the Royal Mail, and the General Register Office for Scotland (GROS) has further harmonized and improved the consistency and compatibility of postcode grid referencing. However, it still does not provide a single national infrastructure of definitive addresses and related property information and mapping. Therefore, in September 2002, four government agencies, the Local Government Information House (LGIH), and the Royal Mail announced a joint program to achieve this purpose, known as the ACACIA project.

Local government has traditionally used external as well as internal sources for their application data. Those OS products that local authorities are entitled to under the terms of the OS/LA Service Level Agreement (SLA) are shown in italics in Box 4.2. This box lists all the products in the OS business portfolio for 2002. Since then, OS Street View (ideal for detailed, street-level display and analysis), 1:25,000 Scale Colour Raster (for environmental analysis), and Points of Interest (a database of location-based information) have been added to the list. In addition to the OS's expanding range of products, the main government sources are the ONS or the GROS for socioeconomic data, the British Geological Survey (BGS) for geological data, and Her Majesty's Land Registry or the Registers of Scotland for land-ownership data. The ONS was formed in April 1966 from the merger of the Central Statistical Office and the Office of Population Censuses and Surveys to give greater coherence and compatibility to government statistics. Its responsibilities include:

- The organization of the decennial census of population and housing in England and Wales
- The registration of vital events such as births, marriages, and deaths to provide high-quality demographic, social, and medical information and analysis
- The National Online Manpower Information System (NOMIS), which is maintained under contract by the University of Durham and provides subscribers with direct access to official government statistics on population, employment, unemployment, and resources down to the smallest geographical area for which they are available (Masser, 1998).

The 2001 Censuses, in both England and Wales and in Scotland, are the first to use the power of computerized mapping, with the OS providing the digital data underpinning both the operation and the analysis of the results. The data is expected to be more freely and widely available than in the past with much of the output distributed over the Web. The 2001 Census results should be incorporated in ONS's Neighbourhood Statistics service that was launched in February 2001 to assist not only the Social Exclusion Unit's important work on neighborhood renewal but also those who are seeking local solutions to local issues.

4.3 WHAT IS A DATA MODEL AND HOW IS SPATIAL DATA MODELED?

The aim of data modeling is to help our understanding of geographical issues. However, the term *data model* has different meanings in different contexts. In their *Introduction to Geographical Information Systems*, Ian Heywood, Sarah Cornelius, and Steve Carver helpfully split the consideration of spatial data modeling into two parts: the model of spatial form and the model of spatial processes. "The model of spatial form represents the structure and distribution of features in geographical space," whereas "in order to model spatial processes, the interaction between these

Box 4.2   Ordnance Survey Business Portfolio 2002 — Product List

Large-Scale Detailed Mapping

- **OS MasterMap**™ (*Topography*) is the new definite large-scale digital map of Great Britain.
- *Land-Line*® *(1:1,250, 1:2,500, and 1:10,000)* is the original highly detailed, large-scale dataset providing comprehensive coverage of the whole of Great Britain.
- **Superplan Data**® is the most detailed mapping of Great Britain and Ordnance Survey's most successful business-to-business mapping.
- **Superplan plots**® are generated from the same source as Superplan Data and have been designed as valuable on-site tools.
- **Siteplan plots**®/**Siteplan Data**™ have been developed as a cost-effective way of plotting onto convenient A4 map extracts for presentations, legal documents, or for supply to local authorities.
- **Aerial photgraphy** provides high-quality aerial photographs, an integral part of the Ordnance Survey map revision system.
- **Landplan**® is the map of choice for site location, farm or estate management, and identifying land use at 1:10,000 scale.
- *1:10,000 Scale Raster* provides high-resolution detailed mapping.

Historical Mapping

- **Historical mapping** provides high-quality copies of maps from Ordnance Survey's extensive archive.
- **Historical Map Data** is an extensive digital archive of Ordnance Survey paper mapping from the mid-Victorian era onwards.

Small-Scale Mapping

- *1:50,000 Scale Colour Raster* is Ordnance Survey's definite raster product, providing a complete digital view of the popular Landranger® paper map series.
- **1:50,000 Scale Gazetteer** contains around 250,000 names taken from the Landranger map series, providing an excellent reference tool and location finder.
- *1:250,000 Scale Colour Raster* product provides entry-level small-scale backdrop mapping suitable for overlaying with individual business information.
- **Strategi**® provides small-scale digital map data for a variety of backdrop applications.
- **Meridian**™2 is Ordnance Survey's mid-scales digital product offering functional and flexible mapping layers.

Location Mapping

- **MiniScale**® is a small scale product designed for use in desktop graphic applications to provide uncluttered backdrop mapping covering the whole of Great Britain.

Address Referencing

- *ADDRESS-POINT*® is a detailed dataset that uniquely identifies and locates precisely all the postal addresses in Great Britain.
- *Code-Point*®/*Code-Point with polygons* provides Ordnance Survey National Grid references to a resolution of 1 meter for point locations representing postcode units in Great Britain, as well as Irish Grid coordinates for postcodes in Northern Ireland. The polygons provide national boundaries for postcode units in Great Britain.

Boundary Data

- *Boundary-Line*™ is a unique specialist dataset of electoral and administrative boundaries covering the whole of Great Britain.
- **Administrative boundary maps** are defining graphic maps outlining all unitary, local authority, European, and Westminster parliamentary boundaries in Great Britain.

Box 4.2   Ordnance Survey Business Portfolio 2002 — Product List (continued)

- **SABE®** (Seamless Administrative Boundaries of Europe) is the first pan-European boundary dataset at this level of detail.
- **ED-LINE** provides census boundary datasets in two levels of detail, digitized from the 1991 Census planning maps.

Roads

- **OSCAR Asset-Manager®** is Ordnance Survey's definite road dataset of Great Britain for the management of road networks.
- **OSCAR Traffic-Manager®** is Ordnance Survey's definite road dataset of Great Britain for detailed route planning.

Height Data

- **Land-Form PROFILE®** provides a stunning representation of the terrain of Great Britain at 1:10,000.

Note: Products shown in *italics* are available to local authorities through the Service Level Agreement.

Source: From Ordnance Survey (2002) *Ordnance Survey Business Portfolio 2002.* Available online at http://www.ordnancesurvey.gov.uk/businessportfolio>2002/listing.htm (accessed February 17, 2003).

features must be considered" (Heywood et al., 1998). In this section we focus on the modeling of spatial form, while process models will be considered in Section 4.8.

There are two main ways that computers handle and display the basic spatial entities outlined in Section 4.1. These are the raster and vector approaches. The raster data model is the simpler of the two and is based on the division of reality into a regular grid of identically shaped cells called pixels. Each pixel is assigned a single value that represents the attribute of that cell. The area that each cell represents varies from a few square centimeters to several square kilometers. This determines the resolution of the grid. Cells become too big as you zoom in and the scale gets larger. The other main disadvantages are that the images lack the intelligence needed for vector-based GIS, and compression techniques are required to keep storage levels to a manageable size.

The vector data model is similar in operation to children's join-the-dot books. Each point, line, node, polygon, or area is uniquely identified and the relationships among them together along with their attributes are stored in the database. This has the advantage of providing intelligent data, but is costly in both time and manpower. The main disadvantage of the vector model is that as datasets are combined and analyzed, a much greater level of processing is required.

The traditional method of representing the geographic space occupied by spatial data is as a series of data layers. Each layer describes a particular use or a characteristic of the landscape with the geographic space broken down into a series of units or tiles. An alternative method of representing reality in a computer is to consider that space as populated by discrete "objects." For example, a local authority property department may need to map and manage a vast array of assets — buildings, school sites, and so on. Each of these can be regarded as discrete objects with empty space between them. This method, which draws on the methods of object-orientated programming, groups the objects into classes and hierarchies that more accurately reflect the real world, an approach to modeling that should be easier to understand.

At the root of the reengineering of the National Topographic Database to create the Digital National Framework (DNF) is this recognition that the real world is made up of objects rather than the traditional series of points and lines involved in digital mapping. To reflect this object-orientated view, OS has converted all of its 230,000 detailed mapping tiles to the seamless MasterMap data source containing some 416 million features. These features are labeled with 16-digit topographic identifiers (TOIDs) that are like digital hooks onto which any associated data can be hung. They have the potential to link datasets together unambiguously, thereby allowing public agencies to share information on issues such as crime and social indicators.

Most of the earlier GIS took a two-dimensional perspective of the world at a particular point in time. Yet, the features we are trying to model have a third dimension and are often highly dynamic. While the use of computer graphics can simulate the appearance of the third dimension, this is of little more value than a good perspective drawing and has become known as the "two-and-a-half" dimensional (2.5-D) approach. Construction of full three-dimensional models of geographic space is technically much more challenging.

Writing in *GIS: A Computing Perspective*, Michael Worboys (1995) contested that the dynamic dimension had always been the poor relation in GIS despite the fact that both people and objects respond to new circumstances and events by changing their roles, locations, properties, and behaviors. However, during the second half of the 1990s, handling information about time — the temporal dimension — became a hot topic for research and development, and the rapid growth in both location-based services and vehicle navigation services has increased the need for real-time data. Worboys (1995) distinguishes between temporal systems that handle data relating to events at a given point of time in the past, the present, or the future and dynamic systems that are required to be responsive to events as they happen in a rapidly changing and evolving scenario (i.e., real-time systems). For example, a temporal GIS would be required to handle a set of maps depicting changing land use patterns in the last 50 years, whereas a dynamic system would be needed to respond to rapidly changing patterns of traffic in a transportation network.

4.4 WHAT METHODS OF DATA CAPTURE ARE AVAILABLE?

The data-capture requirements are twofold. The first is to provide the physical devices for capturing data external to the system and inputting to the database. The second is to provide software for converting data to make them compatible with the data model of the database and to check the correctness and integrity of data before entry into the system. As system hardware and software become cheaper and provide more functionality, the cost of spatial data capture increasingly dominates and can account for as much as 70% of total GIS costs.

All data collected in analogue form, e.g., paper maps, ledgers, and photographs, need to be converted to digital form by any one of the following methods:

- Keyboard entry, used for attribute data that are available only in paper records
- Manual digitizing, commonly used for capturing features from paper maps
- Scanning, used when raster data are required for producing, for example, background maps
- Automatic line following, appropriate when transferring distinctive lines from a map, such as county boundaries, railway lines, and contours

Whatever method is chosen, data capture is a time-consuming process. Therefore, for collecting up-to-date information on the location of street lights or the boundaries of playing fields or active mineral workings, the process needs to be automated as much as possible through the use of total survey stations, global positioning systems (GPS), and data loggers attached to other scientific monitoring equipment. Of these, the growing trend is toward using GPS as the most efficient and cost-effective way to collect new features and maintain existing data. GPS is a positioning technique using either a constellation of the U.S. Department of Defense satellites or Russia's GLONASS limited-life satellites together with a portable receiver to dynamically determine coordinates. When selective availability — the deliberate degrading of satellite signal accuracy for security reasons — was discontinued by the U.S. in 2001, GPS users saw an improvement in positional accuracy from the 100 m applying previously to 10–20 m. An accuracy of better than 1 m can be obtained by Differential GPS using data from stationary reference receivers in known positions in conjunction with data from a roving GPS field system.

In February 1999, the European Commission announced that it intended to develop Galileo, a nonmilitary GPS. By March 2002, the European transport ministers had agreed on the resources to fund the project's development phase together with the European Space Agency. Galileo should be operational by 2008, using 24 satellites.

The increasing use of GPS in conjunction with GIS has brought more people into contact with the necessary coordinate transformation to relate the GPS coordinates with those of the OS's National Grid. This transformation, introduced in 1997, is now known as OSTN02 and has an accuracy of 10 cm.

As well as GPS, satellite imagery and Light Detection and Ranging (LIDAR) systems are gradually being assimilated into everyday use. LIDAR systems work by sending a laser pulse from an aircraft to the ground and measuring the time taken for the signal to be returned. Its precise position is calculated using an integrated GPS, and it can provide not only surface elevation data accurately, rapidly, and cost effectively even in poor weather conditions but can also measure the height and density of vegetation. LIDAR offers distinct advantages over other techniques in applications such as coastal zone monitoring, flood zone mapping, and the derivation of 3-D city models.

As the World Wide Web expands the range of devices that can tap into databases, it makes sense to have users find data, crunch numbers, or manage business processes via powerful Internet tools such as ESRI ArcIMS. Geographic information is stored at the server side, transferred to users, and displayed at the client side. Fueled by the e-government initiatives, both service providers and users are increasingly requiring spatial data around-the-clock and in a form that readily integrates with other information. The growth of Web-based products has produced an increase in Net-based GIS solutions for the Internet and the corporate intranet. Web mapping, for example, is the concept of displaying, in a Web browser, maps that are generated dynamically by a map server. OS has recognized the importance of this surprisingly simple concept (geographic information is stored at the server side and displayed at the client side) and their vision is to provide an online geo-spatial data warehouse containing the complete range of its products.

4.5 WHAT TYPES OF DATABASES ARE USED IN GIM AND WHY ARE THEY SO IMPORTANT?

According to Worboys (1995), "The database is the foundation of a GIS." It helps to ease the conversion from raw data to information by ordering, reordering, summarizing, and combining datasets to provide the desired output. A database holds not only the basic data but also the connections between that data. In short, a database is a store of interrelated data that can be shared by several users. These data are managed and accessed through a database management system (DBMS), but for a database to be really useful, it must be secure, reliable, correct, and consistent as well as technology proof (see Box 4.3).

There has been a gradual evolution of database models through time from the early tabular databases (e.g., a simple spreadsheet), through the hierarchical and network databases developed in the 1960s, to the relational and object-orientated database models used at the present time. Most work on databases for GIS has been based around the use of the relational model and this is still the most common. Here the data are organized in a series of two-dimensional tables, each of which contains records for one entity. These tables are linked by common data known as keys. Querying these databases can be facilitated by menu systems and icons and by the

Box 4.3   Databases in a Nutshell — A Review of Database Requirements

In order to act effectively as a data store, a computer system must have the confidence of its users. Data owners and depositers must have the confidence that the data will not be used in unauthorised ways (*security*) and that the system has a fail-safe mechanism in case of unforeseen events such as power failure (*reliability*). Both depositers and data users must be assured that as far as possible the data are correct (*integrity*).

There should be sufficient flexibility to give different classes of users different types of access to the store *(user views)*. Not all users will be concerned how the database works and should not be exposed to low-level database mechanisms *(independence)*. Data retrievers will need a flexible method for finding out what is in store *(metadata support)* and for retrieving it according to their requirements and skills *(human–database interaction)*. The database interface should be sufficiently flexible to respond differently to both single-time users with unpredictable and varied requirements and regular users with little variation in their requirements.

Data should be retrieved as quickly as possible *(performance)*. It should be possible for users to link pieces of information together in the database to get added value *(relational database)*. Many users may wish to use the store, maybe at the same data, at the same time *(concurrency)* and this needs to be controlled. Data stores may need to communicate with other stores for access to pieces of information not in their local holding *(distributed systems)*. All this needs to be managed by a complex piece of software *(database management system)*.

Source: From Worboys, M.F. (1995) *GIS: A Computing Perspective,* London: Taylor & Francis.

use of a standard query language (SQL). However, SQL was not really developed to handle geographical concepts such as "near to," "far from," or "connected to" (Heywood et al., 1998).

As early as 1995, Worboys indicated that there are problems with the relational approach to the handling of spatial data. This is because spatial data do not naturally fit into tabular structures, in addition to the limitations of SQL mentioned above. The main alternative is the object-oriented approach. This "arises out of a desire to treat not just the static data-oriented aspect of information, as with the relational model, but also the dynamic behaviour of the systems" (Worboys, 1995). The static aspect of an object is expressed by a collection of its attributes (e.g., its name and size) whereas its dynamic "behavior" is represented by a set of operations (e.g., roads used by children to get from home to school).

Whatever the approach adopted, a key element of database philosophy is data sharing. As the volume of databases held by local authorities expands, the number of users grows, and the need for joined-up thinking increases, the importance of database management becomes even more critical.

4.6 WHY IS DATA QUALITY IMPORTANT AND HOW DO WE ACHIEVE IT?

The AGI (1996) published valuable guidelines on geographic information content and quality. These stressed the importance of ensuring that any data acquired was fit for its intended purpose. The guidelines also highlighted five different aspects of data quality:

- Completeness — the measure of the inclusion or exclusion of items from the database
- Thematic accuracy — the accuracy of the values of attributes
- Temporal accuracy — the accuracy of values of time-related attributes
- Positional accuracy — the accuracy of the values of geographic position
- Logical consistency — the degree of conformance to any rules that apply to an object or between objects

"Fitness for purpose" is a well-worn phrase but nevertheless important. All GIS users should strive for quality products from their systems and aim to produce high-quality output. The old computer saying of "garbage in, garbage out" recognizes that if you put poor quality data in, then poor quality output results. Indeed, any errors in input data are likely to be compounded during GIS analyses, thereby further misleading end-users. Success in using GIS to aid decision making is inextricably linked to the quality of the data used.

Heywood et al. (1998) recognize that there are two issues of particular importance in addressing quality and error issues: (1) the terminology used for describing problems, and (2) the sources, propagation, and management of errors. As it is essential to describe the data quality problems before resolving them, the various terms used are clarified in Box 4.4.

Box 4.4   Describing Data Quality and Error

Problems that affect the quality of individual datasets:
- Error: physical difference between the real world and the GIS facsimile
- Accuracy: the extent to which an estimated data value approaches its true value
- Precision: the recorded level of detail of the data
- Bias: the systematic variation of data from reality

Data quality is also affected by some of the inherent characteristics of the source data and the data models used to represent data in GIS. These include:
- Resolution: describes the smallest feature in a dataset that can be displayed or mapped
- Generalization: the process of simplifying the complexities of the real world to produce scale models and maps

Datasets used for analysis need to be:
- Complete: both spatially (cover the entire study area) and temporally (the time period of interest)
- Compatibility: datasets that can be used together sensibly
- Consistency: datasets developed using similar methods of data capture, storage, manipulation, and editing
- Applicability: describes the appropriateness or suitability of the data for a set of commands, operations, or analyses

Source: Adapted from Heywood, I., Cornelius, S., and Carver, S. (1998) *An Introduction to Geographical Information Systems,* Harlow, U.K.: Longman.

While clarifying the terminology is the first step to providing quality GIS, the next is to examine the possible sources of error. Both spatial and attribute errors can occur at any stage in a GIS project. These include errors in the source data and errors in the data modeling, conversion, analysis, and output stages. Despite considerable research effort, little has been done to incorporate error identification within proprietary GIS packages (Heywood et al., 1998). Errors are, however, a GIS fact of life, but adopting good practice in data capture and analysis by following advice such as that provided by the AGI (1996) should be sufficient to keep errors to a minimum.

4.7 WHAT ANALYSES ARE TYPICALLY CARRIED OUT ON SPATIAL DATA?

Data analysis is a key process in transforming data into information, and there is a wide range of functions available in all GIS packages. Heywood et al. (1998) provide an excellent introduction to this subject and demonstrate that the methods used and the results obtained vary in accordance with whether raster or vector data are used. In this section, we summarize the seven basic functions identified by them and indicate how they might be practically applied in local government:

- Measuring lengths, perimeters, and areas
- Performing queries on a database
- Buffering and neighborhood functions
- Integrating data using overlays
- Interpolating
- Analyzing surfaces
- Analyzing networks

Measuring lengths, perimeters, and areas is probably the most common application of GIS in local government. Virtually every service in local government needs to measure lengths (of roads, footpaths, safe routes to schools, etc.), perimeters (of boundaries), and areas (of buildings, playing fields, planning application sites, etc.), which, if done manually, can be a tedious and time-consuming task. By using vector GIS, not only are these calculations much quicker and usually more accurate but also the lengths and areas data can be stored as attributes in a database and so need to be measured only once.

Performing queries on a database is an essential part of GIS analysis — whether to check the quality of the data input (do all data points representing street lights appear alongside highway?) or to answer questions after analysis has been undertaken (how many primary schools have more than a hundred pupils?). This second example illustrates that queries can be aspatial as well as spatial. Aspatial queries are questions about the attributes of features, in this case the type and size of school, rather than their location. Individual queries are often combined to identify entities in a database that satisfy two or more criteria, for example, "How many residential units have been allowed in the green belt in the last 10 years?" Reclassification of cell values can be used in place of the query function in raster GIS to identify areas of particular importance to the user, e.g., areas liable to flooding.

Buffering is used to identify a zone of interest around an entity. Creating a circular buffer zone around a point to answer the question "How many houses are within 400 m of a proposed incinerator outlet and what are their addresses?" is the easiest of the buffering operations in vector GIS. Creating buffer zones around line and area features is computationally more complex, but essential when analyzing road networks or the impact of large waste disposal sites on the surrounding area.

"The ability to integrate data from two sources using map overlay is perhaps the key GIS analysis function. Using GIS it is possible to take two different thematic map layers of the same area and overlay them one on top of the other to form a new layer. The techniques of GIS map overlay may be likened to sieve mapping, the overlaying of tracing paper maps on a light table" (Heywood et al., 1998). At its most basic, a map overlay can be used for the visual comparison of data layers, e.g., overlaying vector traffic information on a raster map background. On the other hand, overlays can produce new spatial datasets from the merging of two or more layers. For example, selecting the site of a new library will involve investigating a whole range of criteria relating to land use, accessibility, deliveries, and others.

The role of interpolation in GIS is to fill in the gaps between observed data points. A common example is the construction of height contours on topographic maps. GIS packages contain a number of techniques, of which Thiessen polygons, triangulated irregular networks (TINs), and spatial moving averages are the most common. Thiessen polygons assume that the values of unsampled locations are equal to the value of the nearest sampled point. Their most common use is to establish area territories for a set of points, e.g., the construction of areas of interest around population centers. A TIN is a method of constructing a surface from a set of irregularly spaced data points. It is often used to generate digital terrain models (DTMs). The spatial moving average "involves calculating a value for a location based on the range of values attached to neighbouring points that fall within a user-defined range" (Heywood et al., 1998). Examples of suitable applications include the interpolation of census data, questionnaires, and field survey measurements.

DTMs create surfaces for analysis, including the calculation of slopes and aspects. Some GIS packages allow you to "walk" or "fly" through a terrain model to visualize what the view would be like at various points on or above the DTM. This can be enhanced by draping other data onto the surface of a DTM (such as an aerial photograph) to add realism to the view. Digital elevation models (DEMs) are similar to DTMs but include surface features such as buildings and vegetation. This detail is provided by laser scanning and is invaluable when used in applications such as line-of-sight modeling, flood risk analysis, and woodland management.

Finally, network analysis can be used to address classic problems such as identifying the shortest routes for waste collection vehicles and the safest route to the nearest primary school.

Data analysis is an area of continuing development as vendors and academics provide solutions to the growing demands of users. Some software products focus on just one of the functions described above, e.g., network analysis, while others combine several of the methods to improve GIS functionality.

4.8 HOW DO MODELS OF SPATIAL PROCESSES HELP DECISION MAKING?

By simulating the real world, a process model helps us to understand the often complex behavior of physical and human spatial systems. Although these models do not provide answers, they do help us to improve our understanding of a problem and to communicate our ideas to others.

In GIS, three different approaches are used — scale analogue models, conceptual models, and mathematical models. Scale analogue models are scaled down and generalized replicas of reality (Heywood et al., 1998) such as topographical maps and aerial photographs. Conceptual process models express verbally and graphically the interactions between real-world features. The most common conceptual model in GIS is the systems diagram that uses symbols to describe its main components and their linkages, and frequently indicates both inputs and outputs. Figure 4.1 is an example of a conceptual model designed to specify the portfolio of a council's property and those relationships that are important to them in dealing with that property. Mathematical models use a range of techniques to help us understand trends and make predictions or forecasts about the future.

In GIS, the three approaches can be used in isolation or combined into a complex model. Whatever the approach, their aim is to help the user make decisions by providing clear and easily understandable information. For example, they can be used to predict the changes to traffic flows if a new business park were to be given planning permission or if a new section of road were to be built. They can also indicate how the siting of a new supermarket could influence the shopping patterns of both local residents and visitors to the area. In cases like this where both distance and attractiveness are examined, gravity models are often used to compute the relative attractiveness of the related shopping centers.

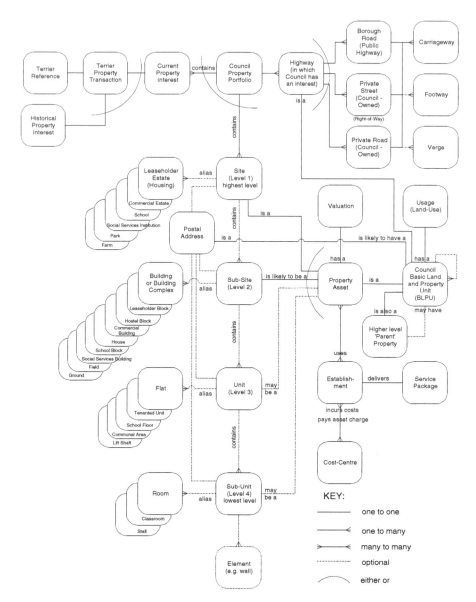

Figure 4.1   Proposed detailed data model for council property. (From Peter Thorpe Consulting, "Council Owned Property Information Project," Report Study to London Borough of Enfield, April 1998. Reproduced with permission from London Borough of Enfield.)

4.9 WHAT ARE THE MAIN FORMS OF GIS OUTPUT?

After capturing data of the right quality, storing it in a database, and analyzing it, the final step in the process of converting raw data into information is to present it to those who are going to use it for decision making and problem solving.

Maps are still the most common form of GIS output and have long been used to support decision making. Most people are fascinated by maps and, as an established part of our culture, they are difficult to beat as a means of visualizing information generated by GIS. What better way is there of identifying the hotspots of crime in a district than a background map overlaid with precisely located points of recorded incidents?

Microsoft's MapPoint and AutoRoute Express products demonstrate the popularity of map-based systems for both analyzing and communicating information. AutoRoute Express is an example of a consumer product that has evolved into a powerful, easy-to-use mapping application bundled with an extraordinary amount of data for very little money. MapPoint combines a rich base map with core spatial functionality such as map rendering, i.e., detailed and easy-to-read maps, enhanced demographic data, proximity searches, tracking, and routing.

As the name implies, the OS's popular *Interactive Atlas of Great Britain* also enables the user to interact with the information stored on the disk and choose between a range of scales and a selection of layers. It also illustrates that GIS packages can provide facilities for the display and playback of multimedia — in this case, some examples of video clips and photographs — to supplement the traditional map.

Aerial photographic images provide much more than a pretty picture. When orthorectified and combined with OS map data, they provide a powerful geographically accurate base from which one can derive new information or update existing databases. These digital orthophoto maps are becoming more popular for a variety of applications from land management to civil engineering design and environmental assessment.

Despite the popularity of maps and photographs, some attribute information is still best presented in tables and charts. Area or ward profiles are a good example of this. Nevertheless, much of the output of the 2001 Census will be provided as high-quality thematic maps to aid the presentation of the data.

Although many users still feel most comfortable with output in the form of paper maps, tables, diagrams, or photographs, an increasing proportion of geographic information is transmitted electronically through e-mail, intranets, and the Internet.

Many local authorities are meeting the challenge of the growing demand for online information. Wandsworth, for example, has been using the Internet for output ever since the council launched its online planning register in 1996. Later, the council commissioned the development of an online planning and building control Web enquiry system. Here the general public can query applications on the database, monitor progress, see the planning constraints affecting the application, examine the listings of all the statutory consultees and neighbors consulted, and link into the drawings on the register. The Website was an immediate success with over 18,000 hits per month recorded in 2001 (Rix, Markham, and Howell, 2001).

Using digital mapping in tandem with automatic vehicle location systems and route tracking data, it is possible not only to monitor traffic flows but also to convey up-to-the minute information about local traffic and available parking spaces to travelers. This is achieved via road traffic broadcasts, roadside messages signs, and bus information displays as well as the World Wide Web. ROMANSE (Road

MANagement System for Europe), first developed in Southampton in 1992, is the best-known example of this, and it is now being extended to Winchester and other parts of Hampshire.

While this section has illustrated some of the main forms of GIS output, there are many others too numerous to mention here. Some other examples are given within the case studies described in Part 3 of this book, and new forms of output are constantly arising from advances in technology. Recent developments in wireless technology, virtual reality, and 3-D visualization have widened the scope for disseminating GIS output. Keeping up-to-date with these latest technology trends is just one of the topics discussed in the next chapter, which focuses on the third and final element or leg of GIS, the technology.

CHAPTER **5**

Technology

KEY QUESTIONS AND ISSUES

- How important is technology to GIS/GIM?
- Why is it important for those involved in GIM to keep up-to-date with new technology?
- How does the diffusion of new technologies take place?
- What are the contents of a local authority's technology kit bag?
- What are the main GIS software products used by local authorities?
- What are the major technology trends affecting GIM?
- How do those with an interest in GIM keep up-to-date with the new technology?

5.1 HOW IMPORTANT IS TECHNOLOGY TO GIS AND GIM?

New technology is revolutionizing our lives and now is the time to harness this revolution and make sure that it takes place for everyone, not just the chosen few. The internet has brought with it mass communications and the ability for any individual to access and contribute to a phenomenal amount of information. We are now in the information age and our Government wants an information age government. This means modern joined-up services providing best value for the citizen (UK Favourites.com, 2000).

Almost all services, whether provided by government or commerce, relate to a place. The motorist using a WAP telephone may want to know where the nearest McDonald's is, the local resident may want to know the opening times of the nearest leisure center and how to get there without a car, and the politician may want a map showing the location of all crimes in a ward in the past year. While the requirements are different, the need is the same — to identify the location concerned and then either to provide information about it or to record a problem relating to it (or its occupants) that requires a response. As local government's responsibilities extend to some 700 different services and as there has been exponential growth in the availability of information about location, i.e., GI, the list of possible questions that could be posed extends toward infinity.

Although many of the basic methods used to handle GI are similar to those employed before the advent of computers, managing the large amounts of spatial data now available to local authorities and displaying them in graphical form would simply not be possible without new technology. Frequently, this is technology that has been developed for other purposes, such as military, medical, or multimedia, and adapted to the needs of the GI industry. The result is an impressive array of products that is expanding all the time to provide one of the three supporting legs of any GIS. Without their contributions, the whole process would literally collapse.

5.2 WHY IS IT IMPORTANT FOR THOSE INVOLVED IN GIM TO KEEP UP-TO-DATE WITH NEW TECHNOLOGY?

The essence of any GI technology is its ability to integrate data from a wide range of sources through the use of a geographic reference. It enables us to perform tasks that were previously accomplished manually much more economically, efficiently, and effectively. More importantly, it also enables us to undertake new tasks that were not previously possible. However, as GIS are such sexy, seductive technologies, it is easy to become captivated by their glamour and wizardry. Therefore, it is always important to remember that any computer system employed in local government should exist to serve both human and organizational needs. With a growing range of products now labeled GIS, costly mistakes can be made if the wrong one is chosen or if new products are ignored through lack of awareness. It is absolutely essential that the users, not the system vendors, remain in control by asking what the technology will do for them rather than being captivated by demonstrations of what a product has achieved for other users operating in different environments.

During the early years of GIS, the systems were very much in control and too many applications were technology led. Keeping abreast of new technologies was left to the "white-coated" experts. But as we have seen in Chapter 2, this began to change during the 1990s as the systems became more user-friendly and improvements in processing capacity enabled vastly greater quantities of data to be manipulated. By the second half of the decade, an increasing number of users were assuming control, thereby broadening the spread of those needing to keep up-to-date with the available new technology and its potential for their organization. However, for the reasons outlined at the end of Section 2.5, the proportion of local government officers and members actually using GIS has remained small. The result is that most staff are less efficient, many services are less effective, and much policy making is less comprehensive than they could be.

Research undertaken by Heather Campbell and Ian Masser (1995) in the early 1990s found that while technology is highly evocative, it is also notoriously difficult to define. Their view, which we support, is that technology involves people and techniques as well as hardware and software. Just as successful cooking needs a good cook as well as a good oven and a recipe book, there is no IT without people. Using Sproull and Goodman's (1990) definition of technology as the "knowledge of cause-and-effect relationships embedded in machines and methods," Campbell and Masser conclude that all technology contains three common elements: machines, methods, and knowledge. They point out that although common usage tends to associate technology with progress, believing it to have "a near-mystical capacity" to improve the future well-being of the whole of society, it is the response of users to that technology that really makes the difference. This emphasizes the point made by the Chorley Committee of Enquiry that technology is more than just items of equipment, with user awareness critical in determining the take-up of any new technology (DOE, 1987).

Implicit within much of the discussion about technology is the idea that it is new or innovative. In reality, it is generally the machines or the methods that are innovative, while the way in which technology is conceptualized, understood, and used is based on existing knowledge and practices.

In the light of the rapidly expanding technologies, the general lack of awareness within local authorities is worrying because huge amounts of spatial data collected at public expense are not fully utilized in either service delivery or policy making. On the other hand, a growing proportion of commercial organizations, including the utilities, are finding that using up-to-date GIS technology gives them competitive advantage as well as improves efficiency. In addition, the new technologies highlighted in the e-government strategy, e.g., the Internet, digital TV, call centers, and mobile phones, all handle geographic information. As each of these present new ways of improving service delivery, so grows the need for more local government politicians and senior officers to be aware of their potential benefits.

5.3 HOW DOES THE DIFFUSION OF NEW TECHNOLOGIES TAKE PLACE?

Diffusion refers to the process whereby technological innovations such as GIS are adopted and taken up by various user groups. It is a relatively slow process until a critical mass of users is achieved. Once again, Campbell and Masser (1995) have researched the main features of the GIS diffusion process and the extent to which adoption and use of GIS is facilitated or impeded by the institutional and organizational context. Following Campbell's earlier work, they examine three explanatory theories of organizational change (Box 5.1) with respect to the diffusion of GIS: *technological determinism*, where innovations diffuse simply because of their inherent technical advantages over existing practice; *economic determinism*, which regards computerization as an essential prerequisite for economic survival both in the public and private sectors; and *social interactionism*, which assumes that technology is socially constructed and views diffusion as interaction between technology and potential users within a particular cultural and organizational context (Campbell, 1996).

Campbell and Masser (1995) draw three important lessons in relation to the two determinist theories:

- The outcome of diffusion of computer-based technologies is by no means universal.
- It is not the technology itself that determines the level of adoption, but the particular circumstances and institutional contexts in which the system is located.
- The virtual exclusive focus is on equipment and machinery, with little or no consideration given to issues concerning existing knowledge and expectations.

Box 5.1 GIS Diffusion — Theories of Organizational Change

Technological Determinism

This approach assumes that the inherent superiority of new technology means that it will inevitably diffuse. (*If someone develops a better washing powder, it is bound to sell*). Innovations enable old tasks to be undertaken more effectively as well as opening up new areas of activity. Diffusion can be constrained by technical considerations and the skill (or lack of skill) of potential users. However, most literature describes a system's potential based solely on its technical capabilities. "Often articles read as if developing more powerful and user-friendly applications will automatically result in the blossoming of GIS in practice" (Innes and Simpson, 1993, quoted in Campbell, 1996).

Economic Determinism

Here the emphasis is on computerization as the essential next step in economic development. It is assumed that technological advances in production and communication will increase prosperity and even the quality of work experience — and consequently the general well being of the whole society. On the other hand it is recognized that introducing new technology will cause some problems of adjustment within the organization. The conception of an organization, however, is still that of a rational, almost machine-like, structure which is amenable to logical adjustment. The belief is that problems caused by the introduction of new technology can be accommodated by logical restructuring of organizational procedures, possibly using business process reengineering methodology.

Social Interactionism

Here organizations are viewed as very complex social structures, which cannot be expected to behave rationally. Organizations are viewed as being composed of groups of individuals, each with their own motivations and ambitions. In such a conception, the adoption of new technology, no matter how impressive, is by no means assured. Whether an information system is a success will depend upon complex interaction of, often informal, political and social forces within the host organization. Technology becomes more widely diffused if it becomes fashionable. If the people inside an organization cannot be persuaded to adopt a new system, the system is little more than very expensive junk.

Sources: From Campbell, H.J. and Masser, I. (1995) *GIS and Organisations: How effective are GIS in practice?* London: Taylor & Francis; Campbell, H.J. (1996) Theoretical perspectives on the diffusion of GIS technologies, in Masser, I., Campbell, H., and Craglia, M. (Eds.) *GIS Diffusion: The Adoption and Use of Geographical Information Systems in Local Government in Europe,* London: Taylor & Francis, pp 23–45; and Reeve, D.E. and Petch, J.R. (1999) *GIS Organizations and People: A Socio-Technical Approach,* London: Taylor & Francis.

These suggest that diffusion is far more complicated than the linear view often presented. Despite the inherent elegance and attractiveness of the arguments underlying the determinist positions, they are unlikely to work in the real world of local government.

Although lacking a certain elegance, the social interactionist approach does seem to offer more in terms of accounting for the varied experiences of individuals and organizations in practice. Campbell and Masser believe that the widespread diffusion of GIS depends on the acceptance of the technology merits within each organization, particularly its ability to cope with change and its approach to decision making. To consider the diffusion of technology as a linear (inevitable) process is not only misleading, it can also distort expectations.

Historically, GIS in local government has been largely decentralized and bottom-up in nature. It has often been championed by enthusiastic middle managers in key departments, such as planning, highways, or estates, and often geared initially to meeting the authority's mapping needs. The expectation was that once a local authority had adopted the technology in one department or for one application, it would be "rolled out" to other services where its benefits would be realized. But this has frequently not been the case, partly because much of the early hype was not realized in practice, partly because it has been perceived as a niche technology developed separately from the authority's other IT activities, and partly because of a lack of awareness and commitment of politicians and senior managers. As a result, GIS diffusion in most authorities has been slow and its corporate impact limited.

5.4 WHAT ARE THE CONTENTS OF A LOCAL AUTHORITY'S TECHNOLOGY KIT BAG?

Although the diffusion of GIS within local government has been disappointingly slow, most authorities have a sizeable technology kit bag developed over the last 30 years largely for accounting, recording, and communication purposes. As a result, most officers and many members regularly use desktop or laptop computers in their daily activities. While they are not computer experts, they are familiar with using Microsoft Windows and mouse-based interfaces, carrying out simple queries on datasets, producing reports by word processing, and sending internal e-mails. For example, a structured survey of professional planners in 1999 revealed that 93% had access to a computer at work, and in nearly 70% of the cases this was for the respondent's personal use (RTPI, 1999).

The list of items in a local authority's technology kit bag is illustrated in Box 5.2. This list is getting longer as new technology becomes available, uses of it become more widespread and user friendly, and awareness of what is available grows. Until recently, a major limitation of desktop computers was their difficulty in handling the large volumes of data required for a GIS. This is now being addressed through use of client/server networks, increased use of databases and data warehousing, and Web-based products.

The most striking feature of two surveys undertaken by the Royal Town Planning Institute (RTPI) in terms of the type of hardware platform used for GIS is the

---

Box 5.2   The Technology Kit Bag

- Hardware/peripherals/networks (including printers, plotters, digitizers, scanners, data loggers)
- Programming languages (e.g. C++, Visual Basic, Java)
- Databases (e.g., MS Access, Census, NOMIS)
- Processing systems (e.g., planning applications)
- Analysis (e.g., statistics, models, projections)
- Tools (word processing, e.g., MS Word; spreadsheets, e.g., MS Excel; computer-aided design)
- GIS (ESRI, MapInfo, Autodesk, Cadcorp, GGP, Innogistic)
- Communication (e-mail, Internet, video conferencing, desktop publishing, presentation packages e.g., MS PowerPoint)
- Recent developments (e.g., WAP, GPS, laser range-finders)

Box 5.3   Application Service Providers

The wider use of the Internet has encouraged several companies (Oracle, Microsoft, and Autodesk among them) to become application service providers (ASPs) that sell a service which guarantees access to their programs and data downloaded over the Internet rather than delivering their products in packages that typically contain a manual and some discs. In the May 2000 edition of *GI News,* Bob Barr concludes that rather than treating software as a capital purchase, it becomes a utility like a telephone service or a power supply (Barr, 2000). As a result, costs are more predictable and the treadmill of "upgrades" unnecessary.

ASPs allow users to access data via the Web and run Web-based applications to process the data. Software is provided on demand via the Web from remote servers. Users do not have to download and install it on their PCs, resulting in economy-of-scale benefits. Also users have access to potentially sophisticated geospatial analysis tools on a pay-per-use basis, thus avoiding the cost of buying, installing, and maintaining the software locally as well as having ready access to up-to-date data and programs.

dramatic growth of the networked PC between 1995 and 2000 — a rise from 50% to 79% at the expense of UNIX and stand-alone PCs (RTPI, 2000).

The term *enterprise computing* has grown in popularity in recent years. It is used to describe the situation where all the users of an organization have access to a central information resource. In GIS terms, this might mean the vast majority of users operating desktop GIS to query a central dataset over a network. The central database would be maintained and updated by specialists using high-end toolkits. Enterprise GIS develop when spatial data are used as an integrated organization-wide resource with all information systems supporting the operation, thereby giving greater efficiency and effectiveness over departmental systems.

Hardware, software, and development costs can still be substantial though various options are now available. On one side of the spectrum is the dedicated, developed GIS application on your own PC, which will require an individual GIS software licence for each workstation with regular maintenance and upgrade costs. At the other side is the Internet-driven application that offers a standard set of GIS functions on any workstation that has a Web browser. When using the Internet, you have the option to develop and run the application yourself or run it as a service from an application service provider (see Box 5.3).

Just like any individual or organization handling technology, knowing when to upgrade or acquire new equipment raises important resource implications for a local authority as well as questions about whether to rent or buy. These are questions that can only be answered as part of the selection and implementation process considered in the next chapter when both costs and benefits are analyzed.

5.5 WHAT ARE THE MAIN GIS SOFTWARE PRODUCTS USED BY LOCAL AUTHORITIES?

This is one of the most frequently asked questions by those authorities wishing to embark on a GIS or develop an existing one. Indeed, there are several well-known GIS suppliers who regularly supply software to local authorities, for which the market leaders are ESRI (best known for their ArcGIS integrated family of products)

Box 5.4   Examples of Products Used by Local Authorities

Company	Local Authority	Product/Use
Aligned Assets	Reading and Cardiff	Mapping solutions
Assist Applications	Greenwich LB	Axis 2000 GIS
Autodesk	Shepway*	MapGuide — interactive themed maps
BHA Cromwell House	Hackney LB	Data services
By Design	Pembrokeshire and Newport	Corporate mapping tools
CDR Group	North Lincolnshire	MapInfo GMIS — grounds maintenance
	Cambridgeshire	MapInfo TARS — road traffic accidents
Cadcorp	Medway	Active Server Component — Web-based GIS
Data Insight Ltd	Reigate and Banstead	PlannerSearch — constraints and history
GGP	Oxford City	PC-based GIS
	East Ayrshire	NLPG GIS — land/property gazetteers
Geoinformation Group	Over 160 LAs	Cities Revealed aerial photos
Geowise Ltd	Aylesbury Vale*	ArcView Print Wizard
ICL	Rochdale	GIS MapViewer with fast access
	Maidstone	GIS Local Land Charges
Landmark Information	Newcastle upon Tyne*	Historical data — for contaminated land
Philip's	Reigate and Banstead	Philip's/OS street-level data
SIA Ltd	LB Newham	dataMAP SMART Education GIS
Sological Solutions Ltd	Buckinghamshire	Data enhancement service
UK Perspectives	Kent	Precision aerial photographic dataset
XYZ Digital Map Co.	Edinburgh	1:10,000 city map data

* Case study authorities considered in Part 3 of this book.
Source: From *GIS 2000 Event Guide.*

and MapInfo. A visit to a major GIS exhibition or a look through the AGI's *Source Book* will demonstrate that there are a large number of other companies offering GIS-related software products to local government, e.g., Autodesk, Cadcorp, Innogistic, and GGP.

As an indication of the GIS software products available, we reviewed the list of those exhibitors offering products suitable for local government use at the GIS 2000. This annual event organized by CMP in association with the AGI conference was held at Earl's Court in London and is described in the *GIS 2000 Event Guide* as the biggest GIS event in Europe (CMP, 2000). Just over two thirds of the 120 exhibitors listed were offering products suitable for local authority use. These products include digital data and mapping solutions, targeted GIS applications, and data capturing services in addition GIS software solutions. The examples given in Box 5.4 are purely a list of those specifically mentioned in the *Event Guide* and so, while they give an indication of the range of products available, they in no way provide a representative sample of local authority use.

A further indication of software products used in local government is gained by looking at examples of products commissioned by individual local authorities. The list in Box 5.5 is compiled by selecting a range of examples quoted in the product news pages of the technical press (in particular *GI News, Mapping Awareness,* and *GEO Europe*) for a 2-year period commencing in April 1999. Both Boxes 5.4 and 5.5 include some of those authorities comprising the case studies analyzed in part 3 of this book, and these are indicated by an asterisk.

Box 5.5   Random Selection of Software Products Used by Local Authorities
(1999–2001)

Product	Local Authority	Purpose
Cadac	Huntingdonshire	Contaminated land
Cadcorp SIS	North Norfolk	Planning, land charges, electoral registration
	Poole	Corporate-wide GIS
	Brighton and Hove	Spatial analysis of some 600 datasets
Cities Revealed	Wirral and Scarborough	Coastal revealed aerial survey data
ER Mapper	Camden, West Dorset, and Hertfordshire	Aerial photography for historic mapping
ERDAS IMAGINE	Lancaster City	Geoimaging for coastal defences
GGP System	Salford	Recording street lights/illuminated signs
	Plymouth	Corporate digital mapping and database
	Sedgemoor	Cluster analysis of crime records
	City of Nottingham	Street lighting management
Intergraph Geo Media	Hackney and Redbridge	Publishing geodata on the Internet/intranet
SIA dataMAP	Babergh and Wealden	Planning/environmental/central services
	Wigan, Bromley, Cumbria, Sunderland, Birmingham	Education GIS for pupil admissions and transport planning
	Richmond	RICHMAP links datasets to land parcels
XYZ Digital Map Co.	Edinburgh	

Sources: From *Mapping Awareness*, *GEO Europe,* and *GI News* from April 1999 to April 2001.

5.6 WHAT ARE THE MAJOR TECHNOLOGY TRENDS AFFECTING GIM?

Technology is improving at an ever-accelerating pace and is now at a stage and at a price that enables a very wide range of applications. In this section we highlight some of those technology trends that have affected, or are likely to affect, GIM in U.K. local government. Box 5.6 provides a summary of these trends broadly grouped into the three main elements of technology: machines, methods, and knowledge.

During the 1980s and early 1990s, most of the technological focus was on improving equipment and machinery. Box 5.6 illustrates that while better machines, especially those facilitating data capture, still feature in the list of technological trends to watch, improvements of methods as well as knowledge and awareness have grown in importance.

As Box 5.7 indicates these trends are converging to create the potential for a revolution in the way we access and use geographic information. This is aided by the growing number of strategic alliances, and partnerships and mergers between the major players, which are now a critical part of modern technical and business solutions. This is because no single company can build complete turnkey end-to-end solutions that span data collection, management, analysis, mapping, and reporting. The ESRI alliance with Oracle and Leica is a good example of how a strategic alliance can work in practice. One consequence of such cooperation is that standards for data exchange and system interoperability are vital (Maguire, 2001).

Interoperability between systems, seen as fundamental to the development of the GI industry, is currently a hot topic. As we found in Chapter 2, the OGC (Open GIS Consortium) seeks to achieve transparent access to the diverse geospatial data

Box 5.6   Technology Trends

Machines

- Continued improvements in computing processing speeds and storage capacity
- Growing number of handheld devices, personal systems controlled by individuals, and increasing miniaturization (e.g., personal digital assistants (PDAs), palmtops, pen-based systems)
- Improved wireless devices and field systems (e.g., WAP-enabled mobile phones)
- Improvements in GPS receivers and remote sensors, particularly RADAR and laser scanning (LIDAR), and high-resolution satellite imagery
- Merging of PDA, wireless, and GPS technologies, providing the ability to quickly verify information and ensuring data are correct in real time

Methods

- Analysis and 3-D visualization techniques — multidimensional (3-D/4-D) viewing, virtual reality, data search, and integration tools
- Improved image compression
- Data integration supported by improved standards
- More instrumentation to provide real-time monitoring and aid vehicle-navigation, traffic monitoring, weather and pollution monitoring, etc.
- Transactional updates becoming a major data source
- Personal systems/spatial locator, mobile location services
- Pace of Web mapping increasing rapidly
- GPS accuracy improved as intentional degradation is stopped
- Rapidly developing interfaces between GIS and technologies such as visualization, database management and data warehousing, the Internet, and real-time information manipulation
- Greater "plug and play" technology making applications simpler to use
- Widespread use of the intranet — "the vehicle for realizing the dream of GIS for all" (Coote, 1997)
- Virtual reality tools used to create synthetic world representations

Knowledge

- Pervasiveness of technology — intrusion of spatial technologies in our daily lives (multimedia, computer games like SIM City, Digital TV, WAP-enabled mobile phones)
- Consumer solutions, e.g., Microsoft MapPoint 2000, via the World Wide Web)
- Increased spatial literacy, ability, and geo-understanding
- Instrumentation of the environment — a major source of real-time data
- More user partnerships and collaboration between stakeholders
- Growing citizen involvement in government through local empowerment

and geo-processing resources on a networked environment by providing a suite of open interface specifications. To a considerable extent, the increasing integration between GIS and the Internet owes its extraordinary growth to open industry standards from bodies like the OGC and ISO.

The Internet, with its potential to connect virtually every computer in the world, makes database technology more crucial than ever. Most operational GIS are interfaced to standard database management systems (DBMS) such as Oracle, Ingres, or Informix, rather than being based upon their own proprietary internal database. In addition, these DBMSs are becoming more and more able to manage raster as well as vector data (Stoter, 2000). It is the growing need to manage increasingly more complex data types, including virtual reality worlds, LIDAR, and high-resolution airborne imagery that is encouraging database technology to move from relational

---

Box 5.7   A Confluence of Technologies Leading to a Digital World

In a release dated 23rd June 2000 and directed at the ESRI International User conference attendees, (the then) US vice-president Al Gore stated that a confluence of technologies — such as accurate GPS, high resolution remote sensing, cheap storage, wireless devices, the World Wide Web, high-speed networks, and open standards — is creating the potential for a revolution in the way we access and use geographic information. Al Gore, who strongly believes in the importance of GIS, has called for the creation of a "Digital World." This implies a digital representation of the planet that will allow people to explore and interact with vast amounts of information. He is therefore delighted that ESRI, the National Geographic Society, and many other organizations are working together to build a Geography Network.

And a Geography Network

The Geography Network is a collaborative, multi-participant system for publishing, sharing, and using geographic information on the Internet. Jack Dangermond, president and founder of ESRI, stated that because GIS is location-based by nature, it has the potential to become the primary organizing factor in an increasingly complex and interrelated world. He was therefore proud to announce the advent of the Geography Network. In his opening address to ESRI's International User's Conference, Mr. Dangermond said: "This is the most exciting thing that we have ever done! It is a new platform for GIS. It offers new ways to cooperate in the development and sharing of information, provides a portal for spatial data cataloguing, and connects users with the data they need."

Source: From *GIM International,* vol. 14, no. 8, 2000.

management (RDBMS) toward object-orientated (O-O) and component-based developments (Stoter, 2000).

The OS has reengineered its National Topographic Database to introduce an object-orientated approach to storing and retrieving its data. This is a major component in the development of the Digital National Framework (DNF), which aims to provide a step-change in the ways that the OS's customers and partners can access and use geospatial information. The DNF provides a consistent and maintained national base against which anyone's geospatial information can be referenced, either through National Grid coordinates or through unique identifiers. These topographic identifiers, or TOIDs, are numbers which will be given to every individual feature of the landscape — from buildings to fields — mapped out by OS (Ordnance Survey, 2001, and Tyrrell, 2001).

The growing GIS–Internet convergence is also extending into mobile phone/WAP technology. WAP, which stands for wireless application protocol, is a means of transferring data to mobile devices. It is described by the WAP Forum (www.wapforum.com) as "an open global specification that empowers mobile users with wireless devices to access easily and interact with information and services instantly." This means that it is a set of standards that allows content providers to provide information in a format that can be received by users of mobile devices. With WAP-enabled mobile phones now available in the U.K., and browsers available for palmtops, the market for WAP content is now set to grow dramatically. The most common use so far is the "Where's the nearest…?" functionality. With the development of wireless and mobile applications, anyone will soon be able to measure, view, and edit geospatial data at any time at any place and for multiple purposes.

The pace of Web mapping innovation has increased rapidly over the millennium as a result of OGC efforts and has stimulated the emergence of:

- Geography Markup Language (GML), which is the geographic data extension to the Extensible Markup Language (XML)
- The Open GIS Web Map Server (WMS) specification, which specifies the request and responses protocols for open Web-based map client/server interaction
- The growing importance of location-based services and application service providers

Raster data is in the ascendancy. The widespread availability of compressed imagery and its ability to be viewed by a variety of GIS programs, combined with the increasing power-to-price ratio of desktop computers, enables digital imagery to be used for numerous applications and across all industries — including local authorities like Kent and Dudley (Denniss, 2000). While visualizations are useful in helping us understand the visual character of the real world, they are currently falling short in the area of analysis and in their ability to utilize multiuser enterprise databases.

The integration of surveying and GIS components is bringing the worlds of 3-D measurement and 3-D GIS together. This promises to accelerate progress in building and managing high-quality 3-D databases (Maguire, 2001). In fact, Leica Geosystems (with its investment in Cyra Technologies) has taken a further step towards providing users with full 3-D data acquisition tools.

Finally, knowledge and awareness of technology is growing as more people use global positioning systems, talk on WAP-enabled mobile phones, view digital TV, send e-mails, and surf the Net. Computer games like SIM City and products like Microsoft's MapPoint, while not groundbreaking in terms of GIS functionality, give more users practical experience in handling geographic information.

5.7 HOW DO THOSE WITH AN INTEREST IN GIM KEEP UP-TO-DATE WITH THE NEW TECHNOLOGY?

Keeping up-to-date with all IT developments is a time-consuming activity, so it is probably better for users to keep abreast of those developments which are being discussed in GIS circles by using a combination of the following options:

- AGI annual conference and exhibition, source book, and regular AGI newsletters
- User group conferences organized by vendors, e.g., Autodesk, ESRI, MapInfo
- Other conferences, e.g., Digital Mapping Show, the World of Surveying, RTPI conference "IT & GIS for Planners," AGI Local Authority Interest Group conferences, GIS Research U.K. annual conference
- Networking through Digital Mapping Connection — www.digitalmappingshow.com (networking event for those in the north of England and Scotland), user groups, or the Society of IT Managers (in Local Government) (SOCITM)
- Web sites of the major players, e.g., www.ordsvy.gov.uk or www.esri.com
- Technical magazines and journals e.g., *GI News, GIM International, Geo Connexion* (previously *GEOEurope* and incorporating *Mapping Awareness*)
- Professional journals and publications, e.g., *Geomatics* (formerly *Surveying*) *World, Planning, Geo Informatics*

As keeping up-to-date becomes increasingly difficult, more local authorities are keeping abreast of the major technological trends with the help of GIS consultants so that they can focus their own GIS expertise on first establishing the needs of their authority and then making sure it is provided. This leads us neatly into the next chapter, which investigates the approaches to GIS justification, selection, and implementation.

CHAPTER 6

Approaches to GIS Justification, Selection, and Implementation

KEY QUESTIONS AND ISSUES

- How do you get GIS "on the starting blocks"?
- Why is it important to have a strategy?
- How do you make the business case?
- What factors should be considered when selecting a GIS for a local authority? Which are the most important and why?
- What is the role of cost-benefit analysis in the implementation process?
- What are the main ingredients of successful implementation?
- Why is a user-centered approach required for successful implementation? What are the training needs?
- How important are continuous monitoring and review?
- What problems are likely to be faced by organizations implementing GIS?

6.1 HOW DO YOU GET GIS "ON THE STARTING BLOCKS"?

In the opening chapter we recognized that many GIS implementations have gone wrong because organizations were not quite sure how they should be used. Therefore, having spent the last three chapters describing the three main elements or legs of GIS, we now focus on how GIS should be justified, selected, and implemented. While we agree with Stephen Gill (1996) that there is no single correct or prescriptive approach to introducing GIS into a local authority, there are some important messages to convey about getting GIS on the starting blocks.

Roodzand (2000) notes that "as with all new ideas and the introduction of new technology, it often takes one or two 'believers' in the organization to get it going." These are usually people who combine knowledge of the organization and its processes with an innovative character, a keen interest in modern IT, and an urge to move forward.

Changing the way that GIM is perceived, particularly at the senior management and the political levels of an authority, is essential if GIS are to play a stronger role in the management of local authorities. To invest in GIS, authorities must have a management culture that is both proactive and change-orientated and regards information as an important asset.

As stated by Gault and Peutherer (1990), "GIS are too simplistically associated with map handling, routine 'technical' activities (e.g., engineering, land use planning) and routine land and property management issues. They are not generally seen as broader decision-support which provide a medium for the more efficient integration of disparate data sets or as integral components of strategic planning and management procedures."

These simplistic associations immediately limit many peoples' perceptions as to whether GIS are relevant to their particular interests. As a result, the subject tends to be approached on the basis of answering the question "What can I use GIS for?" rather than asking the more profound question "What sorts of information systems and analytical procedures do I need to manage my business more effectively and efficiently?"

According to Roodzand (2000), "The time spent in creating the right attitude as well as the organization's commitment, is time well spent, or putting it more bluntly, a must." In this process the following steps are particularly important:

- Getting a "champion" at senior management level to put GIS on the political agenda
- Ensuring that end users are on board by selling the project internally
- Getting the right people to the right meetings and fine-tuning any presentations to stimulate the desired audience by addressing both operational and management needs
- Obtaining a clear picture about the users of spatial information and their requirements
- Drawing up an inventory of concrete GIS opportunities to support business processes
- Identifying the priority showcase projects for early implementation including those that support today's burning issues and political hot buttons
- Talking business not technology — promoting those opportunities (rather than the new technology) within the organization
- Identifying barriers before they are brought up and investing in winning over skeptics
- Producing a master plan explaining the overall target, the project's scope, the proposed steps, and the expected costs and benefits; above all, being realistic — because no one will invest in dreams

The process of obtaining the necessary commitment for introducing a system is often influenced by the status and respect that the promoting department or champion possesses within the authority. The project manager is also a key player in guiding the evaluation and selection of GIS technology. This person must have depth of knowledge and breadth of experience, especially in implementation activities. Above all, the project manager must be good at promoting the potential of GIS within the organization. One should not underestimate the challenge that the project could face if it is not sold effectively within the organization at the outset. While each individual will have a different perspective, any GIS development should focus on providing both the viewer and user community with the relevant data for their jobs.

6.2 WHY IS IT IMPORTANT TO HAVE A STRATEGY?

In the past, too many GIS developments in local government have been either technology-led or narrowly centered on a specific process or task rather than focusing on the business needs of the organization. Unless a GIS project has an agreed upon mission, clear objectives, and defined performance indicators, it is impossible for a local authority to know whether it has been successfully implemented.

While truly corporate GIS units are still rare, even departmental GIS can have far-reaching effects on an authority's structure and culture (Gill, 1996). Whether a top-down or bottom-up approach is adopted depends on the organization's needs, its existing information systems, and the availability of spatial data. What is appropriate in one organization may not be in another. It is essential that each authority produce its own strategy to meet its own needs.

Any GI strategy must have the following qualities:

- It must be business-led, not technology-led, and the focus must be on addressing the needs of the business as a whole, not on what current or even anticipated technology will allow.
- It must be policy-led and not process-led, more concerned with why the authority is doing something and its impact than with examining the minutiae of current administrative processes.
- It is essential to adopt a corporate approach if a degree of synergy is to be realized that will assist the management of change and increase operational efficiency.
- It must be actively endorsed by senior management and politicians and be viewed by them as an integral part of the management process.

The Local Government Management Board (LGMB, 1993) usefully distinguished three aspects of an authority's information strategy:

- The information management (IM) strategy, which sets out the policy and priorities of the organization for the management of information.
- The information systems (IS) strategy, which is a long-term IT directional plan. This strategy is business focused and sets out what should be achieved using IT. It is a demand statement and addresses questions about standards, architecture, and risk assessment. It should provide a framework for all new investments and developments, both small-scale/incremental and large-scale/high impact (Gill, 1996). In short, an IS strategy addresses the issue of what information systems are needed to support the business objectives.
- The information technology (IT) strategy, which is concerned with the delivery of the technology and provides a framework within which applications can be provided to the end user. This is a supply statement that addresses the question of the technological framework needed to deliver the IS strategy.

6.3 HOW DO YOU MAKE THE BUSINESS CASE?

If senior managers and politicians are persuaded to further investigate the opportunities offered by GIS, then the next step is to appoint a GIS project group or working party to carry out an initial assessment and build the business case. This will enable judgments to be made about the required level of investment and the expected benefits.

The elements of the GIS business case are in general similar to those of many information and communication technology (ICT) projects. They answer:

- What is it about? (purpose of the project, description of functionality of application)
- What are the costs? (people, data, and software)
- What is the financial impact? (decrease of costs, increase in income, and improvement of the marketing position)
- How long will it take to develop the system?
- How long will it have to last to gain advantage?
- What is required to make any of the advantages a permanent advantage? (Geldermans and Hoogenboom, 2001)

These six issues help managers and politicians to make an educated decision on the viability of the GIS initiative. While the key to a successful business case lies in being brutally realistic about short-term costs (Autodesk, 2000), it also helps to put short-term investment into perspective with the long-term benefits. In fact a key element of the GIS business case is the expected life-cycle description of the project. Although frequently left out of the equation, both implementation and maintenance must also be part of the business case. Indeed, any benefit from GIS will evaporate over time if not upgraded periodically.

While there many similarities with other ICT projects, there are a number of distinguishing features of GIS that make it difficult to calculate the effort required to realize its full potential. These include:

- Geographic data often involve much larger data capture and conversion costs, and these are frequently underestimated.
- The effect of GIS can be so large that the wishes of users (and consequently the functionality of the system) change each time a new part of the system is delivered, so a stepped (or iterative) approach is best.
- Many GIS applications contain functionality that is rarely used, either because there is no real need for it or because it is too complicated to operate; this can add significantly to the costs, so sorting out the desired functionality from the rest is critical.
- In successful GIS implementations, users have found a multiplicity of ways to increase benefits and thus shorten the payback period.

Derek Reeve and James Petch (1999) subdivide the wide range of activities involved in drawing up the business case into two broad activities: (1) scanning the external environment and (2) internal investigations. They caution that "a danger to be avoided is to focus too quickly upon issues solely within the organization." We agree that much can be learned from the successes (or failures) of other organizations and from contacts with potential hardware, software, and data suppliers. Scanning the external environment can both shorten development times and help avoid repeating costly mistakes.

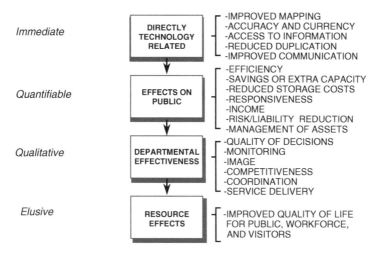

Figure 6.1   The flow-through of benefits from GIS and IT.

Although scanning the external environment will provide a context for the project team's work, Reeve and Petch (1999) emphasize that establishing the internal case for GIS should be the major emphasis of the business case. They believe that this internal investigation should comprise three parts: (1) a user-needs study; (2) a cost-benefit analysis; and (3) a risk analysis.

We have already drawn attention to the critical importance of user requirements analysis in the preparation of the business plan. Indeed, the Geographic Information Steering Group (GISG) of the LGMB (1993) believed the need to determine GIS user requirements to be the most important step in the whole GIS project evaluation life cycle. The task should lead to an understanding of what information is being used, who is using it, and how it is being used. Ideally, it should be conducted from a corporate perspective, and the use of consultants' skills to complement in-house resources can be most valuable at this stage. The flow-through of benefits from GIS and IT is illustrated in Figure 6.1.

Many authors advise that a full cost-benefit analysis should be undertaken as part of building the business case. While it is clearly important to produce a balance sheet of expected costs and anticipated benefits at this stage, we believe that a full cost-benefit assessment cannot be undertaken until after the system has been implemented. Although reasonable estimates can be made of procurement, startup, data capture, and maintenance costs, there is a tendency to make overambitious assumptions about potential benefits. We look at the role of the cost-benefit analysis in more detail in Section 6.5.

Experience with developing GIS leads us to the conclusion that they often do not turn out as planned. Therefore, assessing the risk of failure is an important part of building the business case. Reeve and Petch (1999) recommend using a structured approach to identifying those sources of risk by following the advice of McFarlan (1981) and dividing them into three groups:

- Size risk, i.e., the more man-hours, the more staff, the longer the development time, the more departments involved, the greater the risk
- Structure risk, i.e., the more organizational change required, the more management support demanded, the more job definition changes, the greater the risk
- Technology risk, i.e., the more novel the hardware and software, the less experience the development team has, the greater the risk

6.4 WHAT FACTORS SHOULD BE CONSIDERED WHEN SELECTING A GIS FOR A LOCAL AUTHORITY? WHICH ARE THE MOST IMPORTANT AND WHY?

After putting in place a strategy, establishing what the organization needs, and constructing a business case, the next stage is to select a GIS supplier. This involves several important steps including:

- Invitation to tender specification
- Supplier short-listing and reference site visits
- Benchmark testing
- Project management

The invitation to tender (ITT) is a contractual document setting out the authority's requirements, constraints, and timetable of events together with the contractual conditions. It is the start of a process that involves issuing this document to several suppliers, evaluating their responses to create a short list, followed by on-site presentations and demonstrations, and then benchmark testing. The format and content of the specification is particularly important as it is the basis from which to decide who could supply the solution and how many should be invited to tender. In cases where tenders over €200,000 (about £143,000 in 2003) are likely to be received, the European Union defines mandatory time limits for the receipt of initial "expressions of interest" (37 days) and receipt of subsequent tenders (40 days), which can protract the tendering process. Even where tenders are likely to be under the EU limit, at least 1 month needs to be allowed for suppliers to submit an adequate response. During this period, a tender evaluation panel should be established involving a mix of users ranging from the head of service to the technician.

The ITT document should be structured in a logical and easy to read format and include the following:

- Introduction to set the scene, i.e., background to procurement, description of existing facilities and ongoing strategy, and timetable of the procurement process, including benchmarking
- Overview of requirements
- Detailed statement of requirements (if appropriate, divided into essential and desirable)
- Proposed form of contract
- Format and content of supplier's response

Beware of overspecifying your requirements, thereby leaving the supplier little room to prepare alternative solutions. It is:

- Best to let the supplier propose a system configuration to meet a set of requirements
- Important to allow suppliers sufficient time to respond
- Essential to specify the format of the suppliers' responses in order to ease evaluation

On-site presentations and demonstrations by suppliers have also been successfully used to provide a detailed understanding of the specific GIS solutions that are proposed. These should be followed up by reference visits to those local authorities that are using the systems proposed by the suppliers.

Benchmarking is most important and should not be underresourced. It should be both quantitative and qualitative (user-friendliness, speed, and functionality). If properly undertaken, benchmarking can achieve two important goals:

- An objective assessment of a product's ability to meet a set of predefined requirements
- A subjective assessment of a supplier's capability and commitment

Putting suppliers under the pressures of a benchmark environment provides an insight into the depth of technical experience and the culture of an organization. It also provides a good indication of the flexibility and robustness of the products being tested (LGMB, 1993 and 1993a).

Benchmark tests should be user designed and require potential suppliers to perform a number of prescribed technical tests using the authorities' own data as part of the proposed GIS solution. It is important to involve senior managers from each of the participating departments who have the power to make decisions and those personnel who are skeptical of the operational benefits claimed for GIS.

There may be a temptation to save time and money by missing out the benchmarking stage, especially if the system is small and of limited functionality. However, if not undertaken as part of the evaluation, the purchaser is in danger of investing in a system that has not been adequately demonstrated. It is also important to differentiate benchmarking from sales demonstrations and presentations, which would normally take place earlier in the selection process. During benchmarking, the purchaser, not the vendor, should set the agenda and control the proceedings (LGMB, 1993 and 1993a).

Pilot studies are sometimes used as a means of testing the capacity of a particular GIS within the host organization. Where used, pilot projects are designed to answer the critical question, "Yes, the GIS looks like it will work and can improve performance, but will it work here?" Indeed, a properly designed pilot can provide considerable benefits — demonstrating the potential, helping to gain support, giving hands-on experience, and gaining more of the feel of the vendor company, as well as allowing assumptions about timescales and resources to be tested. However, there are limitations. Reeve and Petch (1999) emphasize that most pilots are not realistic guides as they are almost entirely technical and often flawed because they are based on limited, static test data. Also levels of vendor support can decline once a sale is secured.

Many organizations believe that they derive considerable benefits from pilots, although the term *pilot* is often used when *first phase of implementation* would be more correct. We believe that, in preference to pilot projects, it is better to have a phased approach to GIS implementation with regular reviews.

Speaking at the RTPI GIS conference in 1999, Malcolm Baker stressed the importance of project management by treating the whole evaluation and selection process as one project. He emphasized that evaluations should not focus on just the financial aspects of the project but should also include investigations into the organization, resource levels, personnel skills, training, supplier performance, and alternative technical solutions (Baker, 1999). The evaluation should also review alternative technological solutions on the market and comment on their potential advantages over the current solutions.

6.5 WHAT IS THE ROLE OF THE COST-BENEFIT ANALYSIS IN THE IMPLEMENTATION PROCESS?

The cost-benefit analysis should be designed to answer four key questions:

- What benefits have actually been received from the implementation of new technology?
- Who have those benefits accrued to?
- Are the benefits worth the costs of the IT investment?
- What is the payback period of the IT investment? That is, how long does it take before the cumulative benefits outweigh the IT investment and associated implementation costs?

Other questions on which the cost-benefit analysis will seek to provide a view are:

- What are the disbenefits, additional costs, problems, and unexpected effects that the implementation of technology has caused?
- What do the staff think of the new technology, and has morale improved?
- Are there any areas of lost opportunity where the potential benefits from the GIS have not been adequately realized, and how should these be tackled?

Box 6.1 sets out the main concepts that underlie our recommended approach to cost-benefit analysis. Figure 6.2 provides a conceptual summary of the process of assessing costs and benefits, which also indicates the items of data that will need to be collected. The figure views the organization as a machine that processes a workload from an input of resources (staff and finance) and produces an output at a performance level.

Costs should include hardware and software procurement costs, support costs borne by local authority staff, startup costs (including data capture costs, consultancy costs, and any costs associated with conducting the cost-benefit), and maintenance costs. A common failing is not to cast the net wide enough, and frequently the costs of both management and troubleshooting are excluded, perhaps because both are difficult to quantify in advance.

Box 6.1   The Cost-Benefit Analysis Concept

The major components of the conceptual approach are summarized below:
- Information technology and geographic information systems, with careful implementation and management, can deliver a wide range of benefits to local authorities (see Figure 6.1).
- The potential benefits from IT and GIS may not actually be realized through inadequate management or attention to implementation.
- The benefits that IT and GIS can potentially deliver flow through in a successive manner from benefits that are directly technology related (e.g., improved access to information) and are immediate; to resource effects (e.g., increased efficiency), which are quantifiable; to improvements in departmental effectiveness, which are often qualitative; to ultimate effects on the public (e.g., improved quality of life), which are generally elusive.
- A positive benefit is an improvement in performance actually realized as revealed by a comparison of data collected at two dates (therefore the concept of a baseline date and a subsequent study date which is used in this book).
- An improvement (or reduction) in performance is calculated by identifying a suitable performance indicator and then measuring, and comparing, its values at the two dates.
- For operating (or running) costs, a change in cost is similarly calculated by identifying an appropriate cost category and comparing its values at two dates (within the cost-benefit analysis, a reduction in cost can also be treated interchangeably as a positive benefit).
- Costs of GIS investment need to be converted to an annual basis (either as an annual lease cost or by determining a "write-off period" for the capital investment) in order to be incorporated in the cost-benefit analysis together with operating cost.

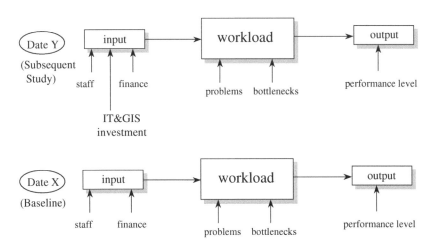

Figure 6.2   The organization as an input–output machine.

Most local authorities do not undertake a cost-benefit assessment as part of implementing new technology. Those that do generally undertake a cost-benefit analysis prior to a decision to invest. The analysis is used to inform the decision of whether to proceed and to provide the basis for bidding for the necessary funds. When it is not feasible to undertake a full cost-benefit analysis, it is still very important to identify any key benefits, even if these are only qualitative. Indeed, many of the benefits relate to better decisions, greater coordination of activities, and improved customer focus, and so cannot be quantified.

Even fewer authorities carry out a review of the costs and benefits actually accruing once the system has been implemented. Again, this review is important so that authorities can assess whether an improvement in performance can be directly attributed to investment in GIS. We believe that it is fundamentally important to measure resource inputs (staff and finance) and workload at regular intervals, as changes in performance may also be partly or wholly attributed to changes in these other factors.

Indeed, we recommend that, as a precursor to an ongoing cost-benefit review, a study is carried out in order to measure the authority's current operating costs and performance levels in those functional areas that will be affected by the implementation of GIS. This is important in order to set the baseline before the operation of departments and services is changed through the implementation of new technology. It provides a yardstick against which improvements in performance can be measured. A further significant advantage in undertaking a baseline assessment is that it enables local authorities to set target performance improvements in advance. It can also be treated as a learning and refining process in order to finalize the approach and "fix" the data items, sources, and methods of collection to be used.

Ideally, the cost-benefit review should be carried out once the systems are fully operational and then repeated on a regular (6-month or annual) basis. It is important to note that the first phase (the baseline study) is in many ways exploratory as it breaks new ground in the field of assessment of cost and benefits — the core of an approach that should be reviewed and refined as the work progresses.

6.6 WHAT ARE THE MAIN INGREDIENTS OF SUCCESSFUL IMPLEMENTATION?

Box 6.2 sets out the six dimensions that we believe are critical to the success of a GIS. The starting point to achieving these six dimensions should be an audit of the authority undertaken by filling in a questionnaire completed either by one of the senior managers or, preferably, by a wider group such as a GIS steering committee. Some local authorities have used a scoring system in this respect, enabling the results to be presented graphically as a performance wheel that enables areas of weakness

---

Box 6.2   Six Dimensions Critical to Future Success of GIS

Vision — The clarity of the organization's aims for using GIS in support of its key business objectives, and the extent of its plans to translate strategy into action
Commitment — The strength of the organization's backing for the introduction of GIS, especially from members and senior managers, and the level of resources, which are pledged
Organization — How well the organization has established working structures, and allocated clear responsibilities for steering, implementing, and monitoring the introduction of GIS
Staff — The extent of support from user staff for the plans for GIS, and the level of skills currently within the organization and related training and awareness programs, which are in place to maximize return on investment
Systems — The degree to which the key requirements have been defined, so as to provide a clear focus for the selection of systems
Data — The extent of the organization's proposals for managing data as a key resource (standards, quality, ownership) and the assessment of the tasks of data capture and conversion

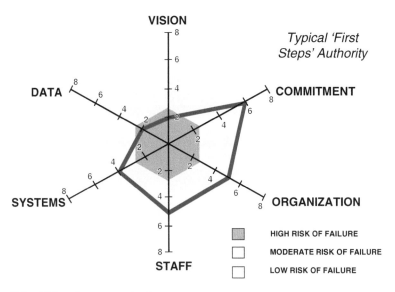

Figure 6.3   GIS performance wheel.

to be quickly highlighted and used as bases for planning corrective actions. The results of the audit may also be incorporated into training and awareness seminars or used to brainstorm the major issues. A performance wheel for a typical "first steps" authority is shown in Figure 6.3.

In assessing success or failure, it is important to keep the boundaries as wide as possible — e.g., a GIS that is successful in its own terms may have caused massive organizational disruption. Therefore, we recommend using any measures of success and failure with extreme caution.

In addition to the six dimensions listed in Box 6.2, Gill (1996) has noted many other ways to improve significantly the chances of success, including:

- Obtaining sponsorship at the highest level throughout the process
- Adopting an innovative, risk-taking, learning approach
- Identifying a GIS champion with enthusiasm, understanding, and commitment
- Appointing a project manager with "bite and presence"
- Earmarking sufficient resources for implementation and data capture
- Obtaining IT support and commitment
- Establishing a multidisciplinary GIS team of problem-solving, lateral thinkers
- Ensuring user representation at all stages
- Preparing a sound business plan with clear aims, standards, and targets
- Introducing staff training, support, discussion groups, and expert consultation
- Undertaking regular evaluations of achievements progress, usage, and products; and good communication, quality feedback, and quality circles

The GIS project manager is the most important member of the GIS team and is crucial to managing the forces of change. The project manager is largely responsible for defining the style for the exercise, setting the agenda, driving forward the agreed upon actions, providing ideas, coordinating efforts, dealing efficiently with problems, and locating and directing resources.

The project manager's job is made easier if there is an active GIS steering group or team to share the duties and responsibilities. Some of the key abilities required are problem solving, brainstorming, imagination, and lateral thinking (Gill, 1996).

Finally, we need to recall Campbell and Masser's (1995) three necessary conditions for effective implementation first considered in Section 3.8. These are as follows:

- An information management strategy that identifies the needs of users and takes account of the resources and the values of the organization
- Commitment to and participation in the implementation of the system by individuals at all levels of the organization
- An ability to cope with change

6.7 WHY IS A USER-CENTERED APPROACH REQUIRED FOR SUCCESSFUL IMPLEMENTATION? WHAT ARE THE TRAINING NEEDS?

A user-centered design philosophy starts from the assumption that human and organizational rather than technological issues are most likely to threaten the effective implementation of GIS. However, the support of users should never be taken for granted. It is crucial that users feel they have ultimate ownership of the project if they are to employ the system effectively in their daily activities. Users' acceptance of and commitment to GIS are unlikely to be achieved if they feel excluded from the processes that help shape their daily work.

It is also worth remembering that technological innovation is of little value if the product of such developments proves problematic to implement or is regarded as irrelevant by potential users. Users require much more than token consultation (Campbell and Masser, 1995).

Effective GIS depends on identifying the real and varied needs of users and having a realistic understanding of the role of GI in their work. This takes time and effort to achieve, especially as users do not always know what they want. Although GIS has its champions, both active and passive supporters as well as skeptics and outright critics, in the middle are the silent majority who are open to persuasion.

Heywood, Cornelius, and Carver (1998) believe that involving users throughout the implementation process helps to:

- Specify functionality requirements (thereby avoid building the wrong system)
- Ensure user commitment and cooperation
- Identify areas where organizational change may be required
- Limit resistance from individuals to any new system

Training is one of the most important ways of supplementing the existing resource base of an organization. However, many local authorities fail to see training as a fundamental investment and inadequately budget for it. We have already stressed the importance of raising in-house awareness from an early stage of the project, as well as throughout the GIS feasibility, development, and implementation processes. This should ease the fears about the technology and transition process, educate staff about the potential of GIS, and keep their expectations realistic. Users should be encouraged to take ownership of their own systems, thereby overcoming resistance and fostering a supportive, committed atmosphere (Gill, 1996).

As users are ultimately the key to GIS success, and the changes that affect them are the most difficult to manage, staff training is essential. However, most GIS courses target computer specialists rather than users. Therefore, on-the-job training is generally the best. This can be individually tailored to the application requirements of the end users.

6.8 HOW IMPORTANT ARE CONTINUOUS MONITORING AND REVIEW?

Change and uncertainty are not only inherent parts of local government, but of GIS as well. These systems are never stable as both hardware and software rapidly become outdated. Moreover, as time progresses, user requirements change and defects in the initial design become apparent. In fact, the implementation process never finishes and continuous reevaluation, justification, and revision are necessary.

If the technology and the data are not kept up-to-date, the GIS applications will soon be of little value. Instabilities within the organization, changes in the controlling political party at either local or national levels, and variations in the external environment all have important influences on the continuing implementation of any technology. Indeed, the very essence of implementation is change, and continuous monitoring and review are therefore vital.

6.9 WHAT PROBLEMS ARE LIKELY TO BE FACED BY ORGANIZATIONS IMPLEMENTING GIS?

As a result of reviewing the implementation of GIS both within local government and outside, we believe that the problems most likely to be faced are:

- Lack of a corporate approach — The diverse range of services to be found within a single local authority frequently inhibits the process of gaining commitment for the introduction of a corporate technology such as GIS. This in turn has implications for the subsequent implementation and utilization of the system. Failure to brief elected members and senior managers and communicate to them the opportunities of investing in GIS is a common pitfall.
- Inadequate resources — GIS implementation is a highly resource-intensive activity involving financial costs, staff time, and a variety of skills. Although training is one of the most important ways of supplementing the existing resource base of an organization, it is essential to make a clear case for the resources (both finance and people) as well as not skimping on (ongoing) training. A distinction needs to be made between resources that are required on a temporary basis (which may

still mean for a period of over a year) and those that will be permanently required to support the ongoing operation of the system following implementation.

- Lack of staff commitment — We have already seen that for GIS to be successful, the staff involved must be committed to making it so. If the staff feel threatened by the speed and severity of the changes they are experiencing, they may still be obliged to accept them, but compliance is not the same as commitment. Therefore, rather than adopting a revolutionary approach to GIS implementation, it is better to embrace an evolutionary style. This allows users to realize the benefits at a more acceptable pace, thereby making sure that everyone who is likely to be affected by the system is allowed some input in the early stages. If the rollout of GIS from the project team is done without sufficient thought for those who will use it, then the system's prospects can be irreparably harmed at an early stage.
- Underestimating the time to create the database — Database creation is complex, costly, and time-consuming, and frequently involves the conversion, capture, and keyboard entry of large amounts of data from existing and new sources. It is also vital because GIS belongs to a category of computer applications that only become useful once a database has been completed.
- Attempting to stick too closely to a standard methodology, especially those adopting a linear approach — GIS implementation is not a linear process with a clear start, clear stages, and a clear finish. It is a learning process, and any methodology, unthinkingly applied, will produce problems because its fixed assumptions will not match the realities and complexities of organizational life.
- Technological advances proved harder than expected to realize — It is vital that the organization can sustain the uncertainties associated with the implementation of GIS technologies beyond the initial flush of enthusiasm that results from is the bandwagon effect of vendor hype and securing the go-ahead to purchase the equipment (Campbell and Masser, 1995).

Facing up to these problems can often be helped by working with others or learning from their experiences. The next chapter picks up on the theme of cooperation and coordination.

CHAPTER 7

Coordinating Mechanisms

KEY QUESTIONS AND ISSUES

- What are the main coordinating mechanisms, and why are they important to GIM development?
- To what extent does the AGI fulfill the role of a national center for GI as envisaged by the Chorley report?
- What coordinating role has central government had?
- What is the IDeA, and what is the focus of its involvement in GIM?
- What is the particular coordinating significance of the NLPG, and what is the role of the NLIS and other "N-initiatives?"
- Why was the OGC established, and what are its main achievements?
- How significant are the European initiatives to GIM development in U.K. local authorities?
- What coordination takes place among GI users in local government?

7.1 WHAT ARE THE MAIN COORDINATING MECHANISMS, AND WHY ARE THEY IMPORTANT TO GIM DEVELOPMENT?

As noted in the Chorley report, "The full potential of Geographic Information Systems will only be secured by coordinating the interest of the widely dispersed community of users" (DOE, 1987). As there was then no single body with responsibility to secure this coordination, the Chorley Committee recommended setting up a central body, independent of government, "to provide a focus and forum for common interest groups in the geographic information area, undertake promotional activities and review progress and submit proposals for developing national policy." For convenience, this proposed body was labeled the Centre for Geographic Information (CGI).

As the management of GI has developed since then so, too, have attempts to coordinate the activity. Indeed, since the publication of the Chorley report in 1987 a number of these initiatives have come and gone with varying degrees of success. For most of that time the chief coordinating mechanism for the GI industry has been the AGI. When it was first established, most of its projects were associated with promoting the use of GI and new systems. The main exception to this was the AGI's work on standards for GI that has been undertaken since it was first established. As understanding developed and a body of practitioners emerged, the AGI has addressed professional development and particular GIM problems.

Other organizations have contributed to emerging national coordination. Professional bodies, in particular the Royal Institution of Chartered Surveyors (RICS) and the Royal Town Planning Institute (RTPI), recognized that geographical information management was part of the professional development of their members. In addition, the RICS provided a home for the AGI for many years.

The representative bodies of the utilities and local government have also made a significant contribution to the development of the industry. The National Joint Utilities Group (NJUG) led much of the early work on the standards for digitizing OS maps, along with the features to be included. The local government bodies, such as the Local Authorities' Management Services and Computer Committee (LAMSAC) under David Hughes, its successor the Local Government Management Board (LGMB) under Tony Black, and the Improvement and Development Agency for Local Government (IDeA) (the LGMB's successor) under Andrew Larner, have developed standards and implemented projects that have a direct benefit to the industry and the citizen alike. David Hughes started the projects that helped local authorities to implement GIS and began the standards activity of what was eventually to become the NSG and the NLPG. Tony Black was involved in the development of the first agreement with OS for the collective procurement of digital mapping (the Local Government SLA) — a development that helped to secure the financial viability of the digitization program. More recently, the IDeA has led the implementation of the NLPG and, with its partner HM Land Registry (HMLR), the implementation of the NLIS.

While many users and providers were involved in this early national coordination work, apart from the acceleration of the OS's digital mapping program, central government's initial response to the Chorley report was both bland and disappointing (see Section 2.3). However, since the establishment of the Inter-Governmental Group on Geographic Information (IGGI) in 1993 the position has improved. Section 7.3 below outlines the contribution that central government has made to coordinate efforts.

There have also been attempts to coordinate GIM activity at the local level. Local authorities have large amounts of geographically referenced information and need coordination to get the most from it. This is because many different people in an authority are involved in managing GI, but the prime purpose of that information can vary significantly between services. This complexity leads to significant differences in the way that information is referenced even within the same authority. This, in turn, leads to very practical problems both for the authority's politicians and senior managers and for their customers.

Where better geographical information management has been achieved at the local level, there has often been a senior champion for change, or a particular project that has had topical significance both to the authority and its customers (e.g., Local Agenda 21).

During the mid-1990s, international pressure from GI providers and users led to the formation of the Open GIS Consortium (OGC) to address key issues like interoperability and better data access, while in Europe several initiatives were having an impact on GIM development. The significance of these to U.K. local authorities is examined in Sections 7.6 and 7.7.

7.2 TO WHAT EXTENT DOES THE AGI FULFILL THE ROLE OF A NATIONAL CENTER FOR GI AS ENVISAGED BY THE CHORLEY REPORT?

Although central government's response to the Chorley report did not result in the setting up of the recommended national CGI, the disappointing response did provide the catalyst for the GI community to coordinate its activities and establish the AGI as an independent organization. Indeed, the AGI's greatest strength is that it represents the whole GI community (data providers, system and hardware vendors, users, public and private sectors) and, with the benefit of hindsight, its independence has perhaps served the community better than anything government-sponsored action could have achieved (Heywood, 1997).

Since 1989 the AGI has not only provided a focus and a forum for both users and vendors, but has also championed and developed many of the recommendations of the Chorley Committee. Its stated mission is "to maximize the use of geographic information for the benefit of the citizen, good governance, and commerce" (AGI, 2002). The AGI operates as a self-help organization that allows people from many backgrounds to come together to share knowledge and experience. It has an ambitious agenda of meetings, seminars, conferences, and publications and members are kept informed through regular newsletters, e-mails, and postings.

The AGI now has an impressive range of special interest groups, including one devoted specifically to local government whose mission is to provide a forum for practitioners to learn from each other in utilizing GI. In 2002 the Local Government Special Interest Group (LGSIG) launched a Website at www.agi.org.uk/lgsig to give details of news and forthcoming events, a library of LGSIG papers, and a growing collection of members' details and case studies to help with networking. Other AGI special interest groups of relevance to local government activities include address geography, crime and disorder, environment, marine and coastal zone management, property, and schools. In addition, each year the AGI organizes a VIP seminar for senior officials from central and local governments. This takes place during the annual conference.

Toward the end of the 1990s the AGI began a number of new devolution initiatives to provide more regional focus in its campaign to demonstrate the value of using GI in all areas of work, recreation, and culture. The first of these, AGI Cymru, was officially launched in Cardiff on November 18, 1999, with the objective of raising GI awareness to people and organizations within Wales. This was followed a week later by the launch of AGI Scotland and then in March 2002 by AGI Northern Ireland. These regional bodies, based in their respective regions, are open to anybody and any organization that uses spatial or geographic information, which in fact is almost everybody. They are each managed by a steering group representing the major GI sectors. In England the AGI/RGS London Initiative brought together the AGI's desire to demonstrate to a wide public and to strategic decision makers the value of modern GI analytical tools, and the desire of the Royal Geographical Society (RGS) to demonstrate the value of these tools in the teaching of geography and in engaging the excitement of pupils in geographic learning. It built on public interest in the election of London's first new-style mayor and of members of the Greater London Authority (GLA) in May 2000. It also aimed to ensure that the GLA was aware of the uses already being made of GI in London and the GI tools available to fulfill their own strategic functions. It has since been developed by the AGI into an ongoing project under the title of the London Information Service.

Many AGI members work in central government, allowing access to decision makers at the highest level. In 2000 the AGI was invited by the Parliamentary IT Committee (PITCOM) to give a presentation on the role that GI has to play both in the information age and in joined-up government. Using the above London Initiative as a case study, the AGI also took the opportunity to highlight the valuable work of both IGGI (see Section 7.3) and IDeA (see Section 7.4). The AGI is also active in Europe, being a founder member of the European Umbrella Organization for Geographic Information (EUROGI; see Section 7.7).

With the achievements of the AGI being so many and so varied, it would be easy to focus on the high profile roles while ignoring their continuing essential work on the development of standards, including the complete revision of the four parts of BS7666 — Spatial Datasets for Geographic Referencing. The AGI Standards Committee is an external committee of the British Standards Institute (BSI), and there are technical panels dealing with data administration, digital imagery, location-based services and system interoperability. Described by Lord Chorley (1997) as "a huge, complex, messy and boring subject, but a vital subject," there is growing awareness that standards are an essential component of interoperability that in turn ensures the effective use of GI.

7.3 WHAT COORDINATING ROLE HAS CENTRAL GOVERNMENT HAD?

The Chorley report (DOE, 1987) recognized that as the biggest user and the largest single supplier of GI, central government and its agencies have an essential part to play in realizing the benefits brought by the wider use of geographic information. However, for several years after the publication of the report there were few visible signs of central government initiatives. Essentially, they were limited to central government being a sponsor member of AGI, and establishing the Tradable Information Initiative (TII). Regarding the latter, the Treasury "ensured that this was a non-starter by removing any incentive for a department to earn a few shillings by selling its data (though they always deny this)" (Chorley, 1997). Nevertheless, encouraged by the progress of the AGI and the persistence of champions like Alan Oliver, IGGI was formed in 1993. Its declared aim was to:

- Provide a forum for government departments and agencies to consider and develop a common view on issues affecting or affected by GI
- Facilitate the effective use, both within and outside government, of GI held by government departments
- Consider barriers to realizing the fullest potential use of government-held GI and taking practical steps to overcome them (DOE, 1997)

Important progress with respect to the availability of data was made following a series of joint AGI/IGGI round-table discussions in 1995 when the barriers to the private sector use of governmental GI were examined (Heywood, 1997). In parallel with these round-table discussions was the development of a Spatial Information Enquiry Service (SINES), a metadata service informing potential users about the availability of some 600 geographically referenced government datasets, and providing the first tentative steps towards a National Land Information Service (see Section 7.5). However, Heywood (1997) believes that the two most important initiatives to emerge from the AGI/IGGI program of discussions were the National Geospatial Data Framework (NGDF) and the GI Charter Standard Statement. The NGDF was a joint initiative by the public and private sectors, launched in 1995, that sought to facilitate the widespread use of geospatial data that is "fit for purpose." It built on the earlier work undertaken by the OS on the development of SINES and had four key drivers:

- Knowledge about what information exists
- Access to information
- Integration of information
- Widespread use of information (Owen, 1999)

The GI Charter Standard Statement, first produced in March 1997 and republished in September 2000, was an integral part of the NGDF. It aims to encourage and promote the effective use of GI held by government by setting out the standards of good practice on information handling that users should expect. Compliance with the statement (summarized in Box 7.1) should in turn assist organizations to meet their own commitments within the information age government program, leading to more joined-up government (IGGI, 2000). IGGI followed up the publication of the Charter Standard Statement by providing general guidance on managing data as a valued resource in a booklet entitled *The Principles of Good Data Management* (IGGI, 2000a).

The need for joined-up working and improved data management is reinforced by Audit Scotland's report *Common Data, Common Sense,* which also stressed the importance of undertaking information audits, agreeing data standards, and appointing data custodians (see Box 7.2).

Following the general election in 1997 it soon became clear that the new government wanted to exploit the power of information and communication technology (ICT) in order to improve the accessibility, quality, and cost effectiveness of public services. The e-government strategy set out the prime minister's vision of a modernized efficient government using the latest developments in e-business to meet the needs of both citizen and business. The government demonstrated its commitment

Box 7.1   Geographic Information: A Charter Standard Statement

Intragovernmental Group on Geographic Information

Government organizations should adopt high standards of data management to support their responsibilities as custodians of public data.
In particular they should:

- Consult users when preparing specifications or drafting legislation for the collection of data.
- Adopt a systematic approach to data management within their organizations.
- Provide up-to-date, accurate information about what data are available and their characteristics, ideally making the information accessible through the National Geospatial Data Framework Data Locator (now www.Gigateway.org.uk).
- Provide clear statements on the charging policy for data and where appropriate the price of that data.
- Make data available, unless there are specific reasons for not doing so, in which case those reasons should be explained.
- Ensure that data adhere to British, European, and international standards and classifications unless there are specific reasons for not doing so, in which case those reasons should be explained.
- Deliver data in standard digital (preferred) or other formats wherever possible to meet users needs.
- Supply accompanying documentation with data to enable users to judge the fitness for any particular purpose.
- Set retention and archiving policy for data, taking into account guidance from the Public Records Office and in consultation with users.
- Publish a contact point to deal with enquiries.
- Provide information about how users can complain if they are dissatisfied with the service they provide.
- Accept responsibility for data quality and publish acceptance of liability in the data they provide.

Source: From Department of the Environment, Transport and the Regions, 2000.

Box 7.2   Extract from Audit Scotland's *Common Data, Common Sense*

"Councils rely on information communications technology (ICT) to underpin their services. Currently, similar data is held many times by different departments. The number of databases and the duplication of records make it more difficult for councils to keep them up to date. Duplication of data also means that overhead costs are higher than they need to be.
Councils are increasingly moving to center services on specific client groups (e.g., combining social work children's services with education). Reconfiguring services involves changes in business processes that need to be supported by integrated information systems. Such joined-up working should involve councils looking at ways in which they can share data — both internally and with partner agencies.
Fundamentally this will mean a culture shift from departments collecting and managing their own datasets to working with others who use similar data. To enable data to be shared, councils need to establish:

- *Data standards* — to make sure that data is held in a standard format which by applying appropriate technical standards, allows it to be shared between different departments and other users
- *Data custodians* — who are allocated responsibility for ensuring that the data held is accurate and up-to-date

Putting these foundations into place leads to a number of further opportunities. These include *customer accounts,* enabling all information about a customer of the council to be linked and be accessible from one point. In the longer term, these steps help councils and their partners to deliver joined-up services."

Source: Extract from Audit Scotland (2000),*Common Data, Common Sense: Modernising Information Management in Councils*, Edinburgh: Audit Scotland.

to a demanding agenda for change by allocating substantial resources to this modernizing government concept.

A Central Local Information Age Government Concordat was signed between local and central governments on July 29, 1999. This committed local government to playing its part in the modernizing government program by pledging to develop the data, network, and legal framework to support electronic service delivery. As the Concordat's program is considered to be best practice, conforming to national standards, particularly BS7666, is seen as an essential element. In fact, Section 7.5 demonstrates that the NLPG is the foundation of all electronic services delivery that relate to land and property. Also the NLIS is perceived as being the first truly cross-government service to be delivered under the Concordat agreement.

Speaking at the GIS 2000 conference, Lionel Elliott estimated that over 40 government departments and other organizations produced geospatial data for their own needs, spending approximately £4000 million annually on data collection and assembly (Elliot, 2000). Until 2000 much of this data was not available for reuse and was often poorly documented. He explained that the key to unlocking this vital information was through the NGDF initiative already mentioned above. This would provide knowledge about what information exists — data about data — and easier integration of the different data sources through a family of sources called askGIraffe operated initially through the OS. The results of the NGDF projects were the askGIraffe Data Locator (launched July 2000) and the askGIraffe Data Integrator (launched September 2000). In September 2001 the NGDF transferred the management and operation of the askGIraffe services to the AGI. In 2002 a new team began work on improving the performance and technical functionality of the service, establishing a brand new name — GIgateway — and a new Website.

The year 2002 also saw the launch of the pilot agreement between the Office of the Deputy Prime Minister (ODPM) and the OS for the supply of OS data across central government. Known as the Ordnance Survey Great Britain Pilot Pan-Government Agreement, this pilot has improved awareness and provided the impetus for much greater use of GI within government. During 2002 it was reported that the number of central government organizations taking OS data had doubled and that a number of citizen services and tools for government based on GI were in the process of being produced (Turner, 2002). Those listed that are of relevance to local government include:

- The Planning Portal, developed initially by the Planning Inspectorate, as the first point of call for anyone with an interest in planning and in finding out about how change in the environment is managed
- The Home Office's JUPITER (Joining Up the Partnerships in the East Midlands), which is an ambitious project giving Crime and Disorder Reduction Partnerships the ability to exchange information
- ONS's Neighbourhood Statistics, which is a Web-based system for small geographical areas, disseminating information on population, access to services, community well-being, crime, economic deprivation, housing, and physical environment, etc.
- DEFRA's MAGIC (Multi-Agency Geographic Information for the Countryside), which brings together rural and countryside information

- ODPM's Maps on Tap, an intranet map-browser service in response to the growing demand from policy makers for accurate, accessible, up-to-date information about what is happening where
- The powerofgeography.com, a joint venture among AGI, IGGI, and OS to promote the benefits of geographic information

In an interview with *GI News*, Vanessa Lawrence, director general of the OS, said that many local authorities have expressed their delight at the pan-government agreement as it now allows them to share information on a common base. "Our new licensing arrangements have freed up a lot of local authority information. Many councils have, for example, used our new Internet freedoms to post information on their websites" (Lawrence, 2002).

7.4 WHAT IS THE IDeA, AND WHAT IS THE FOCUS OF ITS INVOLVEMENT IN GIM?

The IDeA exists to help local government meet the current agenda for change, in response to initiatives generated by central government and those introduced by local government itself. It operates on the belief that:

- Real and lasting change is best achieved by working closely with local authorities and helping them to develop their own capacity to make effective change.
- Wherever possible, its work should be carried out in close partnership with others, including the Local Government Association, the government, the private sector, the voluntary sector, trade unions, the Audit Commission, and regional bodies.

The IDeA aims to:

- Help local authorities improve the way they work
- Help local authorities develop the skills of those working in local government
- Recognize the future challenges facing local government and to respond to those challenges on behalf of local government
- Disseminate local government's achievements at a national level (www.idea.gov.uk)

The IDeA came into being on April 1, 1999, as a company limited by guarantee that is wholly owned by the Local Government Association. Its board of directors is made up of elected members, leading academics, trade unionists, industry figures, and a representative of central government. Working closely with local councils in the way that IDeA does gives it insight into the big issues in the local government such as electronic service delivery, social inclusion, best value, and community leadership. These issues, when taken together, are radically changing the way in which local authorities are making policy, procuring services, transforming their business, and brokering new partnerships.

Since its inception in 1999, the IDeA has established the following services:

- Local Government Improvement Programme (LGIP — a peer-review program)
- Beacon Councils Scheme (a scheme set up to disseminate best practice in service delivery across local government)

- Best value (advisory services)
- Performance support (a scheme to help build the capacity of weaker local authorities)
- Member development (a training program for members)
- Management development (a training program for members)
- Workforce development
- Consulting

Through the LGIP, the Beacon Councils Scheme, and its work on best value, the agency has acquired knowledge on the state of development of local authorities and emergent best practice. Also many of the IDeA's 150 employees have years of experience in local government, while others have been brought in from elsewhere in the public and private sectors, e.g., from Whitehall, the BBC, the Audit Commission, the consultancy firm KPMG, and PricewaterhouseCoopers.

The agency's unit, the IAPU, has led the way in delivering accessible, seamless, and customer-focused e-government initiatives. Some of these initiatives deliver services to local authorities (such as the IDeA marketplace creating an online one-stop-shop for local authority procurement needs), while others deliver services to local government's customers such as the NLIS, providing one-stop access to land and property information. While the most obvious parts of the IAPU's work are the large service delivery projects, these projects are a part of a wider activity to put in place the infrastructure for electronic service delivery by local government.

Under the terms of the Concordat, the IDeA is tasked to monitor local authorities' participation and report progress to the central government. By mid-2001 (and updated by October 2002) all councils in England had to prepare implementing e-government (IEG) statements setting out how they plan to deliver services across the Internet, over the telephone, and face-to-face in peoples' homes or at one-stop shops.

Through the IAPU and its trading arm known as the Local Government Information House, the IDeA is a major player in driving forward the NLIS/NLPG initiatives that are discussed in the next section.

7.5 WHAT IS THE PARTICULAR COORDINATING SIGNIFICANCE OF THE NLPG, AND WHAT IS THE ROLE OF THE NLIS AND OTHER "N-INITIATIVES"?

Since the 1970s, the implementation of GIS solutions has brought about the need for standards so that locations can be uniquely identified. Typically, different departments and organizations have held information in their own sets of pigeon-holes, and this has made data exchange very difficult. The desire for every property to have one Unique Property Reference Number (UPRN) goes back to the DOE's GISP report published in 1972 (see Section 2.2). Although the idea of developing a National Land and Property Gazetteer was taken up by the LGMB in the late 1980s, it was not until the IDeA joined forces with Intelligent Addressing, a subsidiary of Property Intelligence plc, that a BS7666-conformant national directory of Basic Land and Property Units (BLPUs), each with a UPRN, appeared likely to come to fruition. This prospect was enhanced by the appointment of Tony Black as operations director of Intelligent Addressing, as he had headed LGMB's GIS unit before returning to work with the OS. Through the sheer hard work and dedication of those involved, this national gazetteer is emerging step-by-step from the creation of Local Land and Property Gazetteers (LLPG) within each district council or unitary authority. The initiative now has its own Website (www.nlpg.org.uk/ezine) to help keep everyone up-to-date, and by January 2003 some 200 local authorities had linked to the NLPG. This means that nearly two thirds of all the properties in England and Wales have an agreed upon address, a grid coordinate, and a spatial extent.

One of the crucial roles of the NLPG is to enable information collected in different systems to be brought together using standard references. The objectives of the gazetteer include the need "to develop definite lists of all land and property in an area" and the "reduction on effort of compiling core information by doing the task once, at source, and then making it available to everyone who needs it — a fundamental principal of information management" (Black, 2000). The particular coordinating significance of the NLPG may vary among authorities, but there are likely to be some common benefits, including improved consistency, reduced duplication, increased efficiency, improved access to information, and increased scope for developing new products and services.

Speaking at the GIS 2001 conference, Michael Nicholson, managing director of Intelligent Addressing, described the NLPG as the fundamental building block of the information age that provided an index to a local authority's applications and enabled more joined-up working across organizations. Nicholson (2001) went on to identify five key initiatives that had been encouraged by the development of the NLPG:

- Implementing e-government plans
- Linking the National Land Uses Database (NLUD) to the gazetteer
- Linking all electoral registers in the U.K. to the NLPG through the Local Authority Secure Electoral Role (LASER) project
- Incorporating the UPRN in the local authorities Asset Management Plans and the revised CON29 Local Land Charge search form
- Making the NLPG the core of the implementation of the single notification of change of residence project

All the BLPUs within the NLPG will be linked to their nearest street, which emphasizes the related importance of the National Street Gazetteer (NSG). This was created under force of legislation in the mid-1990s from the amalgamation of local street gazetteers for each county or unitary authority.

An operational BS7666-compliant LLPG is an essential prerequisite for involvement in the NLIS in England and Wales and the parallel development of the Scottish Land Information Service (ScotLIS) in Scotland. The aim of the NLIS is to streamline all dealings in connection with obtaining information about land and property — achieved by providing links to all the various sources of information that were previously held at disparate and unconnected locations. NLIS/ScotLIS is not a vast central database of land and property data. Rather it is a service allowing access via computer to comprehensive and authoritative spatially related information from any number of sources.

Stemming from the groundbreaking work of Domesday 2000 in the early 1990s, NLIS brought together the OS, the HMLR, the Valuation Office (VO), the LGIH, and a number of other public and private bodies in a Bristol pilot, which is described in more detail in our case study chapter section (see Chapter 9). In 1998 the NLIS project was awarded £2.3 million from the government's Invest to Save Budget. Its purpose was to enable the delivery of NLIS as one of the first examples of joined-up government with an initial emphasis on services that simplified and speeded up the conveyance process. This relies on the electronic delivery of local authority services for Local Land Charges enquiries. ScotLIS, the sister project in Scotland, ran a pilot in the Glasgow area building on the experience of Glasgow City Council that was due in large part to the pioneering work of Rena McAllister. Eventually, this joint central and local government initiative will provide an online one-stop-shop for land and property related information held by many organizations.

Speaking to the RTPIs IT and GIS Panel in 2000, Bruce Yeoman, who worked with Bristol City Council before becoming a consultant to the IDeA, described the NLIS as the father of the N-initiatives. Other initiatives that support of the modernization agenda include the NLUD, which is another example of a national dataset produced from local authority information. It started with the compilation of information about previously developed land (PDL) in response to the government's need to set regional targets for the proportion of new homes to be built on brownfield sites. However, the long-term vision is to establish a complete, consistent, and detailed geographical record of land use in England that is kept up-to-date and delivers information to users that meets their business requirements (Harrison, 2000).

7.6 WHY WAS THE OGC ESTABLISHED, AND WHAT ARE ITS MAIN ACHIEVEMENTS?

Until the mid 1990s most software and hardware vendors appeared keen to lock in their customers and lock out their rivals. Within many organizations competing units battled for survival or supremacy. Open systems were the exception rather than the rule. The OGC was founded in 1994 to address the issues of interoperability, data access, standards, and specifications that were constraining the widespread adoption of GIS. The OGC's approach was to build consensus and promote a vision of open GIS technologies by creating the Open GIS Specification (OGIS), which enables communication between the diverse geoprocessing systems. In 1999 the OGC established a subsidiary in Munich to focus on European interoperability issues, at the same time stressing that this was not another standards organization. Indeed, the OGC aims to enhance the functionality and performance of vendors' products rather than to create uniformity.

The OGC draws its members from an international array of businesses, government agencies, and academic institutions. Although its members include some of the world's leading software and telecommunications companies as well as computer manufacturers and system integrators spanning a wide range of applications, there was initial criticism from some writers and commentators that the OGC focused too much on the U.S. (in 1998, half the membership was from the U.S.) and had an overly strong academic bias (when the European subsidiary was formed, 80% of its members were academics). While OGC participants can be rightly proud of their interoperability achievements, its impact so far on and direct relevance to U.K. local government is only just becoming evident.

By January 2002, the OGC had some 230 members worldwide, 57 of which were from 16 European countries. Writing in the first OGC column in the monthly magazine *GIM International*, Mark Reichardt reported that after modest gains in 1996 and 1997, the rate of Open GIS Specification output and the rate of adoption by vendors accelerated rapidly. "Twenty five vendors now offer 104 software products that complement OpenGIS Specifications. These numbers grow weekly, as the user community demands greater interoperability" (Reichardt, 2002).

The Open GIS Consortium has created the GML with the intention that all GI users will eventually use the same format. It is derived from XML, which is a Web-friendly format. As GML is used for OS MasterMap and by software suppliers like ESRI, MapInfo Corporation, and Cadcorp, local authorities will be increasingly affected by this development.

One of the OGC's ongoing activities is the Web Mapping Testbed, a forum for demonstrating and testing Internet-related GIS software with the objective of making Web access to diverse geo-spatial data transparent and easy.

7.7 HOW SIGNIFICANT ARE THE EUROPEAN INITIATIVES TO GIM DEVELOPMENT IN U.K. LOCAL AUTHORITIES?

The AGI Council is of the opinion that in order to deliver its mission of maximizing the use of geographic information for the benefit of the citizen, good governance, and commerce, it must influence and take part in the wider European debate and initiatives, especially those related to GI, by taking an active role within EUROGI. During the late 1990s EUROGI worked with the European Commission (EC) to develop GI2000, a draft communiqué concerned with the supply of easily identifiable and accessible geographic information. When the EC made the decision in 1999 not to take the communiqué forward, EUROGI (2000) decided to fill the void by publishing its "Towards a Strategy for GI in Europe" consultation paper. The strategy has five main objectives:

- Encouraging greater use of GI in Europe
- Raising awareness of GI and its associated technologies
- Promoting the development of strong national GI associations
- Improving the European GI infrastructure
- Representing European interests in the global spatial infrastructure debate

In 2001 the EC adopted EUROGI's recommendations, which centered on establishing an EU-funded Geographic Information Network in Europe (GINIE). In the same year the EC launched INSPIRE (Infrastructure for Spatial Information in Europe), whose objectives are to make available harmonized sources of geographic information in support of the formulation, implementation, and evaluation of community policies.

Whereas the AGI has gained an enormous amount of credibility from its contributions and the people, like Ian Masser, whom it has put forward for the EUROGI presidency, most local authorities have had little or no direct contact with the organization. They will, however, undoubtedly benefit from the implementation of the EUROGI strategy, albeit indirectly. We suspect the same is true of EuroGeographics, which was established in 2001 by combining CERCO (Comité Européen des Responsables de la Cartographie Officielle) and MEGRIN (Multi-Purpose European Ground-Related Information Network) with the objective of stimulating increased cooperation on GI issues across 37 member states in Europe. The aim is to achieve this objective by working to make the databases held by the various national mapping agencies interoperable and thereby more widely available.

Of greater perceived significance to individual local authorities than any of these pan-European initiatives are the European-funded projects that give practical assistance to the development of GIS. One example of this is the European GIS Expansion (EUROGISE) project that involved Forth Valley GIS and Liverpool City Council along with local authorities in Ireland, Holland, Greece, and Finland. Submitted in July 1996 under the TERRA Programme, the EUROGISE project was selected for European regional development funding and ran for three years ending in December 2000. More details are given in Box 7.3, and a CD-ROM was produced in 2001 to chronicle the achievements of the project whose aim was to develop techniques and share knowledge in local authority GIS applications.

Two further examples help to illustrate the range of GIS applications that have been developed with the help of European funding. In 1999 the EU, through is

Box 7.3   The EUROGISE Project (Author's summary)

EUROGISE was a three-year project (1998 to end 2000) involving six European partners (Forth Valley in Scotland, Liverpool in England, the NASC Consortium of regional authorities in the west of Ireland, Limburg in the Netherlands, Stavroupoli in Greece, and Tampere in Finland).

Funded by the European Regional Development Fund (ERDF), the project had four priority objectives:

1. To promote an integrated, multisectoral approach to spatial planning
2. To use GIS technology as a tool to assist corporate information management
3. To demonstrate positive and measurable benefits for all partners over the duration of the project in terms of improved service delivery and information management
4. To devise new communication links and enhance existing ones between project partners at all levels by developing political and organizational liaison groups

The most important aspect of EUROGISE was the information and its management, analysis, and dissemination. The challenge of the project had been to consolidate the information necessary to each partner for effective spatial planning, to develop "one-stop information access points" and to facilitate the widespread use of GIS technology as a gateway to both the access and analysis of spatial information.

Forth Valley GIS (FVGIS) led the EUROGISE "theme group" on data management and contributed heavily to the group on metadata. The outcomes from the EUROGISE project fundamentally influenced the implementation of Forth Valley GIS, in particular the establishment of clear data management principles and the development of the concept of one-stop-information access points for property information. FVGIS identified that the highest priority was not more functionality or sophisticated applications of GIS, but simple access to the geographic information that was available. It determined that the most efficient and cost-effective approach to provide access to the widest audience possible was through the implementation of intranet GIS.

Tactical Implementation of Telematics Across Intelligent Networks (TITAN) project, invested £1.6 million into hi-ways, a Web-based initiative to improve access to tourism and services by citizens and business in the Scottish Highlands and Islands. Developed in conjunction with four European partners in Ireland, Norway, and Italy, hi-ways provides online information on employment, education and training, tourism and leisure, transport, and local authorities, thereby improving the potential to deliver better services to the public.

The final example is the Road MANagement System for Europe (ROMANSE) that started as a pilot in 1992 and has since expanded to form a much wider demonstration of intelligent transport systems. Part of EUROSCOPE (Efficient Urban Transport Operations Services Cooperation of Port Cities in Europe), ROMANSE is run by a consortium of partners from the public and private sector led by Hampshire County Council and Southampton City Council. The ROMANSE/EUROSCOPE network has now expanded to include Cologne, Piraeus, Rotterdam, Strasbourg, Genoa, Hamburg, and Cork. Using advanced technology, the project provides real-time information to travelers, network managers, and transport providers with a variety of different systems collecting, collating, and disseminating data. As well as providing current information about traffic and road conditions, ROMANSE also aims to change people's travel choices by promoting public transport as a viable alternative to the car.

7.8 WHAT COORDINATION TAKES PLACE AMONG GI USERS IN LOCAL GOVERNMENT?

The underlying theme of the EU-funded projects is one of user collaboration so that everyone gains from sharing information, experience, and knowledge. Lochhead (2000) notes that "as the role of local government changes from one of sole provider of many services to that of key driver of a new system of governance, embracing the complexity of agencies and organizations involved in the provision of services, the demands of IM become critical to efficient and effective working." The Forth Valley partnership is perhaps the best known local government example of coordination of GIS users in different local authorities. Forth Valley GIS (FVGIS) was established by three Scottish unitary councils (Clackmanshire, Falkirk, and Stirling) following local government reorganization in 1996. The joint team provides the three councils with access to GIS skills and services that would have been difficult for each to afford. It has also enabled them to improve their customer focus though the provision of one-stop access points for property information that operate across a local Internet. Because the needs of city governance are diverse and complex, nearby Edinburgh City Council went a step further by forming the Edinburgh Partnerships with other key agencies from the public, private, and voluntary sectors as well as the local community. Their aim was to develop a 5-year city plan by sharing the information and analytical skills of agencies like East of Scotland Water, Edinburgh Chamber of Commerce, Edinburgh and Lothian Tourist Board, Edinburgh Voluntary Organizations Council, Lothian Health, Lothian and Borders police, Napier University, and Scotsman Publications. Their datashare project uses a GIS to enable user-friendly access to a wide range of datasets provided by the growing number partic-ipating service providers (Lochhead, 2000).

Within local government there is a growing amount of coordination between GIS users. Sometimes local authorities within the same county decide to cooperate to their mutual benefit, but more frequently coordinating mechanisms are prompted by the fact that they use the same software supplier, e.g., MapInfo or ESRI, and so can exchange views and information at their regular user group meetings.

Professional institutes like the RTPI and the RICS help to coordinate the GIS activities of their members through the IT and GIS Panel and Geomatics Division, respectively. This is achieved through conferences, meetings, reports, and the Inter-net. Intranets and newsletters are also used by individual authorities to coordinate the work of their GIS users. The case studies include several examples of intranet use, e.g., Shepway (see Chapter 14).

Having now completed our review of the key elements of GIM — perhaps appropriately ending on the themes of cooperation and learning from the experiences of others — we now move forward into what many readers will regard as the main part of the book, i.e., the case studies. Of these, nine are analyzed in Part 3, with Section 18.1 of the book's final chapter drawing together some of their main messages.

PART 3

The Case Studies

CHAPTER 8

Introduction to the Case Studies

KEY QUESTIONS AND ISSUES

- What is the purpose of the case studies in this book?
- How have the case study authorities been selected?
- What styles and applications of GIM are covered by the case studies?
- How has the information for each case study been organized and presented?
- How can I read and use the case studies depending on my background and interests?

8.1 WHAT IS THE PURPOSE OF THE CASE STUDIES IN THIS BOOK?

The previous chapters have looked at the background of GIM, the principles that underpin its application, and the key areas of use within local government. Part 3 of this book contains nine detailed case studies of how individual local authorities have put GIM into practice.

The main purpose of the case studies is to:

- Provide specific examples of how GIM has actually been applied in a local authority setting, an environment where ideals frequently have to be tempered by the reality of political priorities, style, and culture, and restrictions on resources
- Look at GIM in use ("warts and all") from a viewpoint that will be directly familiar to the many readers likely to be working in local government
- Ensure that the theory of GIM is firmly rooted in practice
- Provide grounds for optimism by highlighting that it is possible to achieve considerable success with GIM by a wide variety of routes, despite the path not being easy
- Explore the many human and organizational factors that are key contributors to success and are generally more important to understand than the technical issues
- Enable the lessons from past experiences to be learned to the advantages of other local authorities who will invest their time, money, and other resources in GIM in the future

8.2 HOW HAVE THE CASE STUDY AUTHORITIES BEEN SELECTED?

The case studies have been selected in order to present a mixed bag of examples of GIM in practice that:

- Cover each of the types (except English shire counties and Scottish unitary councils) of local authority
  - English Unitary Council (Bristol City and Southampton City)
  - Metropolitan Borough (Leeds City and Newcastle City)
  - English District Council (Aylesbury Vale and Shepway)
  - London Borough (Enfield and Harrow)
  - Welsh Unitary Council (Powys)
- Cover the range of styles and applications or are important in terms of the history of GIM implementation in local government
- Are distinctive in each case in terms of their approach to GIM
- Offer important lessons to other local authorities

8.3 WHAT STYLES AND APPLICATIONS OF GIM ARE COVERED BY THE CASE STUDIES?

A wide variety of styles and applications of GIM are covered by the case studies. As Table 8.1 shows, the case studies include examples of:

- Each type (except for English shire counties and Scottish unitary councils) of local authority (as identified in Section 8.2)
- The full size range of local authorities from typical districts (e.g., Shepway at 100,000; Aylesbury Vale at 165,000) up to large cities (e.g., Leeds at 715,000; Bristol at 400,000)
- Different states of operation of GIS, from those with a strict single supplier policy (such as Newcastle City, Aylesbury Vale, and Powys) to those with a more relaxed approach that allows several GIS products to coexist side-by-side (such as Bristol City, Leeds City, Shepway, Enfield, and Harrow)
- The main market leaders in terms of GIS suppliers
- Different styles of GIS implementation, from those with a strong corporate approach (such as Bristol City, Newcastle City, Aylesbury Vale, Shepway, and Powys) to those whose approach has been departmental (such as Southampton City and Harrow)
- A range of claims to distinction from pioneers (such as Bristol City with the National Land and Property Gazetteer, Leeds City with LAMIS, Shepway with Web-based GIS) to grassroots authorities (such as Southampton City and Harrow) and those that have used "killer" applications to get things moving (such as Southampton City with education, Newcastle City with Envirocall, and Harrow with drug misuse and crime and disorder)
- A wide variety of focus services and policy areas for GIS application

Table 8.1 Overview of the Case Studies

Chapter Number (Page Number)	Local Authority	Local Authority Type	Population	Current State of Operation of GIS	Main GIS Supplier(s)	Style of GIS Implementation	What Makes This Case Study Distinctive?	Focus Services and Policy Areas for GIS Application
9 (115)	Bristol City Council	English Unitary Council	400,600	Multi-supplier/ authority-wide GIS	ESRI MapInfo FastMap	Corporate	Pilot authority for the National Land and Property Gazetteer that put BS7666 into practice for the National Land Information Service (NLIS)	Land and Property Gazetteer Local land charges Council-owned property
10 (129)	Southampton City Council	English Unitary Council	216,000	Multi-supplier/ authority-wide GIS	MapInfo Intergraph	Departmental	Grassroots approach to GIS, which has been very successfully in the absence of a strong corporate framework	Very wide range of services, with education as the initial "killer" application for use as a showcase to others
11 (139)	Leeds City Council	Metropolitan Borough	715,400	Multi-supplier/ authority-wide GIS	ESRI MapInfo GGP	Semi-corporate	Pioneering authority for ICLs LAMIS system in the 1970s before disillusionment with large corporate projects set in. Reconceived its approach to GIS from the 1990s	Land terrier Local land charges Planning Highways
12 (149)	Newcastle City Council	Metropolitan Borough	270,000	Single- supplier/ authority-wide GIS	ESRI	Corporate	Strong corporate approach together with development of Envirocall (customer relations management) as the "killer" application for use as a demonstrator	Customer relations management Local land charges Planning Grounds Highways Crime and disorder Council-owned property

Table 8.1 Overview of the Case Studies (Continued)

Chapter Number (Page Number)	Local Authority	Local Authority Type	Population	Current State of Operation of GIS	Main GIS Supplier(s)	Style of GIS Implementation	What Makes This Case Study Distinctive?	Focus Services and Policy Areas for GIS Application
13 (159)	Aylesbury Vale District Council	English District Council	165,000	Single-supplier/authority-wide GIS	ESRI	Corporate	Strong corporate approach to implementation of GIS	Local land charges Planning Council-owned property
14 (167)	Shepway District Council	English District Council	100,000	Multi-supplier/authority-wide GIS	ESRI Autodesk	Corporate	One of the first local authorities to implement Web-based GIS within the framework of a strong corporate approach	Planning Building control Environmental health Local land charges Shoreline management Deprivation
15 (175)	London Borough of Enfield	London Borough	265,000	Multi-supplier/authority-wide GIS	Sysdeco/ICL Now replaced by MapInfo	Semi-corporate	First local authority to capture its addresses and property data to BS7666 standards. Has achieved extensive implementation of GIS despite the lack of a fully corporate approach	Land and Property Gazetteer Planning Local land charges Council-owned property Environmental health
16 (183)	London Borough of Harrow	London Borough	220,000	Multi-supplier/authority-wide GIS	GGP SIA MapInfo	Departmental	Bottom-up approach to introducing GIS without a corporate approach based on the burning political issues of drugs misuse and crime and disorder	Drug misuse Crime and disorder Deprivation
17 (191)	Powys County Council	Welsh Unitary Council	126,000	Single-supplier/authority-wide GIS	MapInfo	Corporate	Very strong corporate approach and early recognition that spatial data about disparate locations is critical to a county with an area of 2,000 mi^2	Education Planning Local land charges Highways Countryside access

---

Box 8.1   Case Study Questions and Issues

- Why was the authority chosen as a case study?
- The background: What has the authority done?
- What organization has it set up?
- What does it plan to do in the future?
- What were the positive drivers and success factors for implementing GIS?
- What were the negative factors that threatened success?
- What have been the practical benefits?
- What are the lessons for others?

8.4 HOW HAS THE INFORMATION FOR EACH CASE STUDY BEEN ORGANIZED AND PRESENTED?

To make things easy and comparable, each case study has been written to a standard format that answers the "Case Study Questions and Issues" that are set out in Box 8.1.

Each case study includes, at the start, an "At a Glance" summary that enables the key facts about the local authority to be rapidly assimilated in order to assist the user in identifying those case studies that are of greatest interest and to act as an *aide memoir* after reading.

In terms of the current state of operation of GIS, the following six-way classification was devised in advance of the case studies in order to denote whether the authority had adopted a single-supplier or multi-supplier approach to GIS and to denote how widespread the application of GIS is within the authority:

1. *Single-Supplier/Authority-Wide GIS*
   Only one GIS supplier's product(s) operational within over four departments
   Authorities covered in the case studies: Aylesbury Vale, Newcastle City
2. *Multi-Supplier/Authority-Wide GIS*
   More than one GIS supplier's product(s) operational within over four departments
   Authorities covered in the case studies: Bristol City, Enfield, Harrow, Leeds City,
      Shepway, Southampton
3. *Single-Supplier/Multi-Departmental GIS*
   Only one GIS supplier's product(s) operational within only two or three departments
   Authorities covered in the case studies: Powys
4. *Multi-Supplier/Multi-Departmental GIS*
   More than one GIS supplier's product(s) operational within only two or three
      departments
   Authorities covered in the case studies: none*

---

* It should be noted that within the case studies we did not attempt to select examples of each of these categories in advance, having placed greatest importance upon identifying local authorities that are distinctive in some respect with regard to their implementation of GIS. The classification scheme was therefore employed after the selection had been determined, as part of our approach to analysis. As our selection process was inherently biased towards local authorities that showed significant activity on GIS and had something to offer it was not perhaps surprising that there are no examples of the last three categories in the classification scheme within the case studies.

5. *Single-Supplier/Single Department GIS*
   Only one GIS supplier's product(s) operational within only one department
   Authorities covered in the case studies: none*
6. *Multi-Supplier/Single Department GIS*
   More than one GIS supplier's product(s) operational within only one department
   Authorities covered in the case studies: none*

We have deliberately used the term *authority-wide* rather than *corporate* in this classification scheme as it better describes the extent of implementation of GIS in this context. However in Table 8.1 (where we have defined the style of GIS implementation or level of cross-department working) we have used the term *corporate* as this more appropriately identifies where there is authority-wide commitment to GIS (which may be evident from the existence of a corporate GIS strategy or establishment of a GIS steering group to coordinate implementation and operation).

8.5 HOW CAN I READ AND USE THE CASE STUDIES, DEPENDING ON MY BACKGROUND AND INTERESTS?

There are a number of different ways that the case studies can be read, depending on the reader's background and interests:

- In full and in sequence, which is the recommended approach if the reader has time, because this takes advantage of all the material and the order in which we have placed them to best "tell the tale"
- By being selective when the reader has limited time or wants to dip into the case studies for specific experiences, lessons, and information that are relevant to his or her background, viewpoint, work experience, organization or a pressing problem, e.g., by:
  - Using Table 8.1 to identify relevant case studies, and then reading these initially, reviewing, and if appropriate reading further case studies
  - Identifying those "Case Study Questions and Issues" (Box 8.1) that are of direct interest, and then focusing upon reading the appropriate sections of the relevant case studies

Above all, when reading the case studies, it is suggested that the reader adopt a critical questioning approach in relation to the material presented by continually asking:

- What does this tell me that is new in terms of my knowledge of GIS and how it might be successfully put into practice?
- How does this reinforce or conflict with what I have learned from earlier sections of this book?
- How does this compare with my own experience of GIS and the organizations that I have worked within that have implemented GIS?
- Would this approach have worked better than the approach my organization has adopted so far?

- What experience and lessons are most relevant to me and the organization that I am currently working in?
- How can I help my organization to improve, based upon what I have learned?

We have spent a lot of time selecting the case study authorities and analyzing and presenting the material about them. We hope that you will enjoy reading this section of the book.

CHAPTER 9

Case Study — Bristol City Council

BRISTOL CITY COUNCIL AT A GLANCE

Key Facts

Local authority name: Bristol City Council
Local authority type: Unitary council
Population: 400,600
Current state of operation of GIS: Multi-supplier/Authority-wide GIS
Main GIS products in use: ESRI's ArcView (50+ licenses), ArcLLC/ArcInfo (11 licenses), and ArcIMS, SDE, MapObjects, LPGTools (1 license each); also AutoCAD (50+ licenses), MapInfo (36+ licenses, FastCAD/Map (1 license), ParkMap (28 licenses), and AccMap (4 licenses)
Applications: Map production, gazetteer, local land charges (LLC), development control, engineering, traffic management, parking, lighting, abandoned vehicles, housing, education, and social services
Land and Property Gazetteer status: Fully operational Local Land and Property Gazetteer (LLPG) that was developed initially as the National Land and Property Gazetteer (NLPG) pilot system for the National Land Information Service (NLIS)
GIM/GIS strategy status: Corporate GIM strategy (approved September 1997)
Forum for steering GIS: Geographic information group (GIG) with separate corporate property group for managing property assets and associated geographic information
Staffing for GIS: No central unit — each department makes its own staffing provision for GIS
Contact details: Information services manager (telephone 0117 922 3117)

What Makes Bristol City Council Distinctive?

Bristol City Council's Land and Property Gazetteer (LPG) was the pilot NLPG that put BS7666 into practice for the NLIS. The Bristol pilot project was begun in 1994 and launched operationally in mid-1998 as a collaborative venture among Bristol City Council, HM Land Registry (HMLR), Ordnance Survey (OS), and the Valuation Office (VO). The project focused on streamlining the conveyancing and local land charges process through the sharing of data among 12 different organizations based upon a common unit of land and property. The pilot was so successful that the council received awards from the Association for Geographic Information (AGI) and British Computer Society (BCS). Bristol City Council has capitalized upon the project to its advantage as the impetus to achieving effective implementation of its LLC system, GIS, and LPG. The council was the pioneer organization at the forefront of testing the LLPG concept before the NLPG program was rolled out to other local authorities, and its experience has significant lessons for others.

Key Stages in the Implementation of GIS

> *Stage 1 (1989 to 1993)* — Early steps in GIS starting with use of AutoCAD within the engineering and planning functions for manipulation of OS maps and OSCAR street network data. Development of Bristol's first street gazetteer (linked to digital mapping), which was used initially by the cleansing system, then as the springboard for development of the embryo property database using "wheel bins" data.
>
> *Stage 2 (1994 to 1998)* — Development of business case for LPG and associated implementation of the computerized LLC system, which created a "head of steam" for the city council's lead involvement in the pilot NLPG/NLIS project. Approval of the council GI management strategy in 1997. Live running of the LLC system and integral LPG by mid-1998 as part of the pilot project.
>
> *Stage 3 (1999 to 2002)* — Rollout of Internet-based GIS to all departments modeled on Bristol City Council's LPGWeb (Internet-enabled Land and Property Gazetteer, OS maps, and high-interest spatial data), metadata database (register of datasets and projects compiled through an extensive information audit), and library (of GIS and information management standards and how-to-use advice adopted and prescribed by the council). Voluntary registration of the council's land and property holdings with the HMLR began in 1999.

Positive Drivers and Success Factors for GIS

- Strong champions with vision and understanding of information management principles
- High levels of skills and awareness of data management issues by key staff
- National pilot project that acted as catalyst to achieving Bristol City Council's ambitions
- Engineering function's need for comprehensive street and property data
- Planning function's need for computerized map-based information
- Pressure upon LLC service to improve performance
- Widespread enthusiasm across departments to use the data available within GIS

Problems that Threatened Success

- Major reorganization of Bristol City Council into a unitary authority (April 1996)
- Lack of corporate top-down commitment to GIS
- Poor appreciation of data management issues by middle management

Practical Benefits from GIS

- Improved performance of LLC function
- Reduction in duplicated updating and improvement in data quality and updatedness
- Improved communication with the public (e.g., faulty light reporting and abandoned vehicles systems with GIS interfaces)

- LPGWeb intranet system facilitated coordination between officers and the public on problems and activities at an individual property level
- Improved communication across the council and with the public and external organizations

---

9.1 WHY WAS BRISTOL CITY COUNCIL CHOSEN AS A CASE STUDY?

Individual local authorities have to decide on their positioning in relation to new technology projects. While some are keen to take a leading role and to promote themselves as go-ahead and at the cutting edge, this can often be a lonely, risky, and uncomfortable position to be in. Many local authorities shrink from being at the forefront, deciding deliberately to adopt the safer approach of not being a pioneer and watching and waiting for systems and initiatives to become stable before they consider implementation.

Bristol City Council is an excellent example of the former — a local authority that has been keen to be on the crest of the first wave and to become the pilot site for the NLPG and NLIS. The energy that the council, together with its partners, has devoted to making a success of the project, has been capitalized upon, enabling it to put in place very early the framework for coordinating exchange of information on a land and property basis within the Bristol area. It has also brought kudos to the city by ensuring that it is clearly seen as far sighted, skilled in land and property systems, and able to take a leading role in projects of national significance. Bristol City Council won the AGI award in 1997 for the LPG, and (together with HMLR) won the British Computer Society award for NLIS in 1998. As background to the establishment of the NLIS pilot, and the associated benefits that accrued from it in a local context, this case study has much that is relevant to all local authorities considering the justification for implementing LLPGs.

9.2 THE BACKGROUND — WHAT HAS BRISTOL CITY COUNCIL DONE?

Bristol City is a large unitary council with over 16,000 employees within 5 major departments (central support services; neighborhood and housing services; environment and leisure services; education and lifelong learning, and social services and health). The council is an example of the implementation of a multi-supplier/authority-wide GIS, using the terminology that we introduced in Chapter 8. The council has a relaxed policy of allowing each business function to purchase and use the GIS product that most meets its own requirements.

Products from the following six different GIS software suppliers are in use across the authority:

- **ESRI GIS** — with ArcView (50+ licenses), ArcIMS/MapObjects (1 license each), LPG Tools (1 license), ArcLLC/ArcInfo (11 licenses) used across all departments (for map production, socioeconomic demographics, street lighting, abandoned vehicles, housing management, LLC purposes, and planning application processing)
- **AutoCAD** (50+ licenses) within central support services, neighborhood and housing services, and environment; transport and leisure (for map production, engineering, 3-D modeling, development planning, landscaping, and property management)
- **MapInfo** (36+ licenses) within neighborhood and housing services and environment, transport, and leisure (for map production and general business mapping)
- **FastCAD/Map** (1 license) within environment, transport, and leisure (for map production and general business mapping)
- **ParkMap** (1 license) within environment, transport, and leisure (for mapping of available parking)
- **ArcMap** (4 licenses) within environment, transport, and leisure (for traffic management and planning purposes)

Development and implementation of GIS has taken place over three major stages:

Stage 1 (1989 to 1993) — Over this period the city council took its first formative steps in GIS and began to build its expertise in street and property gazetteers. In 1989 the former city engineer's department (now part of neighborhood and housing services) began to use AutoCAD for engineering purposes and for the display of OS digital maps and OSCAR (road-center line) data.

By 1992, the Bristol Street Gazetteer (which had been developed as a text system in the late 1980s by the research section of the former planning department, now part of environment, transport, and leisure) was linked to digital mapping (again using AutoCAD) in order to support the development of a new cleansing system for the city engineer.

The street gazetteer was subsequently extended to become an embryo property database using the data available in relation to wheel bins, and began to be used by other major corporate computer systems. The planning department started to experiment in the use of GIS for map production, plotting of planning applications, digitizing of local plans, and highways and transport schemes. At the end of 1993, a major impetus to corporate working was put in place with the setting up of a second-tier property and land user group (PLUG), which came unanimously to the conclusion that the prior and overriding need for the future was the implementation of a central land and property referencing system.

Stage 2 (1994 to 1998) — This was a critical period in the history of GIM within the city council and is why it is given major attention within this case study. It was a phase of corporate foundation building during which attention was focused upon developing the business case for creating the LLPG and computerizing the LLC service, while working closely in conjunction with external partners. A major stimulus to establishing the council's infrastructure for holding land and property information came with the coincident publication of BS7666 (see Part 2 of this book — Specification for a Land and Property Gazetteer), coupled with LGMB's invitation to the city council to join the NLIS pilot.

Participation in NLIS was seen as an important means to raise the degree of officer commitment for creating a corporate gazetteer and testing it out in the context of a new LLC system. The whole thrust of NLIS was to:

- Promote the sharing and exchange of land and property information
- Facilitate access to local and central government information
- Enable the conveyancing process (of which LLC formed a major part) to be carried out online

The authority had suffered a succession of cuts in resources over recent years that had placed high priority services under threat. Participation in NLIS was seen as a means to offset this deterioration in services by:

- Improving information management as an ambitious means to upgrade the quality of services
- Drawing the attention of politicians and senior managers to the fact that problems in LPG creation were of national importance and not merely the preoccupation of particular officers
- Providing the prospect of revenue generation through the marketing of local authority data, which might offset a substantial element of the cost of LPG creation and maintenance

Bristol City Council operated within tight financial constraints. It was keen to develop an LPG compliant with BS7666, but members were insistent that this had to be funded through the computerization of the LLC service. As a prerequisite this meant that a business case had to be made that the project would be entirely self-financing from the revenue and savings that it generated. The desire to comply with BS7666 was entirely pragmatic at this stage, in that it would:

- Achieve the best opportunity for stimulating the convergence of existing city council data sets
- Maximize the scope for the exchange of data with other organizations (e.g., via the NLIS), which was considered to be key to successfully generating revenue from the information services and achieve a degree of self-financing

The reason for LGMB inviting the city to join the NLIS was to enable field trials of BS7666 to be carried out in a heavily populated city area. Babtie Consultants undertook an initial study of the council's land and property datasets in August 1994. Using the go-with-the-flow data modeling approach developed by the LGMB, it identified duplicate data stores, processes, and flows in relation to land and property that suggested that potential savings were likely to accrue from implementing the LPG. The study confirmed the intuitive perceptions of Bristol City Council officers that there was considerable waste and inefficiency in the current systems that could be overcome through the use of a corporate gazetteer. The study also confirmed in broad terms, the findings of the earlier Tyne and Wear Joint Information Study in the 1980s. Disappointingly however, it was unable to quantify the benefits that were likely to accrue from a corporate gazetteer. The consultants' report highlighted that of all the council's services, LLC offered the strongest potential gains from computerization and linkage to the corporate gazetteer as it relied heavily upon the integration of land and property information from many different sources.

At that time the LLC service was run entirely using manual systems with total reliance on paper and card index records and not even the use of word processing facilities. As a result, the processing of 8000 to 9000 searches per year was frequently faced with frustrating delays. Despite these handicaps, the LLC office (which contained 12 dedicated and experienced staff) was by 1994 achieving a turnaround that was just quicker than the minimum acceptable level of 10 days. This was due in large part to the active cooperation of the other directorates that provided information for the response (in particular the planning directorate) and represented a significant improvement on the average of 20 days to process a search that was typical of the property boom of the late 1980s. The council was keen to achieve further improvements in processing speed in the future, particularly if this could be self-financing and result in the byproduct of a tested corporate gazetteer. However the city council had been subject to rate-capping for several years and financial controls were stringent. It was clear that a robust business case would need to be put together covering all aspects of income and expenditure, both capital and revenue. This was particularly important in the context of the imminent reorganization of local government, which was to take place in April 1996 with the Avon County Council and six district councils being abolished and replaced by four unitary authorities (of which Bristol City was the largest).

In 1995 the council decided therefore to invite tenders for a consultancy study with the aims of:

- Demonstrating the feasibility (or otherwise) of computerizing the LLC service
- Identifying whether there was a positive case for the use of GIS for LLC processing

TerraQuest Information Management was appointed to undertake the study (which was funded by an increase in search fees of about £3, levied following agreement with the Bristol Law Society). The study examined three configurations for the provision of a computerized LLC system: fully text-based (non-GIS) system, Windows NT GIS-based system, and Unix GIS-based system. The conclusions of the study (which are summarized below with the kind permission of Bristol City Council and TerraQuest Information Management) were that:

- There was a strong business case for computerizing LLC and for each of the three configurations there was a net positive cash accumulation (at 1994 values) over 5 years (of £419,000 for the text-based system, £296,000 for the Windows NT GIS-based system, and £93,000 for the Unix GIS-based system). Achievement of these figures assumed an increase in fees of £20 per search and a staff reduction of between 5 and 7 staff within the existing team of 12.
- Despite the strong business case, the upfront set-up costs that were to be borne in Years 1 and 2 through outright purchase were considerably higher than the council could afford (over £0.5M for each of the configurations, including the cheapest text-based solution). However, if the council considered leasing, then these upfront costs could be reduced and "smoothed" over future years, making leasing an attractive option.

Box 9.1   Geographic Information Management and the Eight Es of Good Local Government

GISs, if properly structured, financed, and supported by effective data management policies, can have a particular bearing on the eight Es of good local government, for example:

Economy: by providing better information on economic opportunities, locations of potential subcontractors and suppliers for incoming firms, facilitating development proposals, localizing cost center management, etc.
Efficiency: by reducing duplication of effort in data collection, increased data sharing, improved timeliness of data
Effectiveness: by providing flexible means of identifying and retrieving relevant and appropriate information
Equity: by facilitating the supply of comparative data for geographic areas
Empowerment: by improving the ease of access to appropriate data for those that need it
Excellence: by facilitating the validation and update of information
Enabling: by improving the scope for data sharing with outside organizations
Environment: by enabling a more holistic approach to problem-solving and policy formulation; greater focus on targeted needs; reduction in paper-based messaging and data storage

Source: From Bristol City Council, "Corporate Strategy for Geographic Information Management," January 1997.

Based on the results of the consultancy study, the council took the decision to finance, through leasing, the implementation of a local land charges system. The authority considered that the potential benefits from a GIS-based system were fundamental to the processing of local land charges searches, and a text-only solution was discounted. Following open tender in 1996, ESRI was appointed to supply the Windows NT GIS-based local land charges system, which included the bespoke development of an integral LPG conforming to BS7666 (as there were no off-the shelf packages available at this time). The additional costs of delivering and maintaining the gazetteer, estimated at about £120,000, were also rolled into the lease.

In September 1997, the council approved a corporate strategy for GIM with the objective of using GIS systems and spatial data to support the "eight Es of good local government" (see Box 9.1). The strategy set out clear principles for the management and sharing of spatial data within the council and introduced mandatory requirements that all new computer systems holding land and property information should interface to the corporate LPG, and all existing systems should be migrated to link to the gazetteer over the next few years.

The LLC system "went live" in April 1998 and was self-financing from the start, based upon an increase in fees and reduction in staff numbers. The overall project costs were £1.25M of which £400,000 was for the data capture contract. An immediate benefit of the new system was a reduction in search times from 10 days to 5 days, coupled with increased confidence in the accuracy and completeness of the search response. The integral BS7666 LPG was developed and tested as part of the NLIS pilot which was launched in July 1998 and which enabled operational problems to be debated and resolved in conjunction with Bristol City Council's partners such as the IDeA, OS, HMLR, and VO (see Box 9.2). The gazetteer contained about 250,000 basic land and property units, each of which was linked to one of about 6,000 streets. As the prototype for the NLIS, the project was able to demonstrate clearly the value of the BS7666 Gazetteer (see Figure 9.1) in providing the

Box 9.2   The Bristol National Land Information Service (NLIS) Pilot Project

The Bristol NLIS Pilot project was launched in Bristol 1st July 1998 by the Lord Mayor, involving HMLR, OS, VO and BCC. This enables a number of solicitors to carry out on-line conveyancing searches and other land and property based searches from data provided by 12 organizations, through sharing of structured information and GIS. (For more details visit the NLIS website at http://www.nlis.org.uk.)

The pilot proved that adherence to standards such as BS7666, and subsequent data matching to a Land & Property Gazetteer (LPG), enables accessibility and promotes sharing of information which can then be displayed graphically through a GIS. Such was the success of the pilot, it gained national awards from AGI and BCS. Following this pilot a Concordat, signed by various Ministers, has stated that all Local Authorities must have an LPG by 2002. These local LPGs will be part of a national LPG (NLPG) which will allow local authorities access to NLIS. These initiatives are also seen as means to achieve modernising local government initiatives from central government such as one-stop shops, call centres, etc.

Bristol regularly receives visitors and requests for presentations and demonstrations from other local authorities and organizations, which increase as recognition of the Pilot Model's application throughout all market sectors grows.

Source: From Bristol City Council, LPG BS7666 Information Audit: A Joined Up GIS Solution, Report to Management Team, September 1999.

framework for unambiguously pulling together information about a property or plot of land — both for local land charges and wider conveyancing purposes across 12 different organizations in the Bristol area that are consulted by local solicitors as a result of a move of house. Based in particular on the work undertaken in Bristol, the NLIS and associated NLPG programs have now been rolled out to all local authorities for implementation. The IDeA is heavily promoting NLIS as the first lead service under the Central Local Government Information Age Concordat.

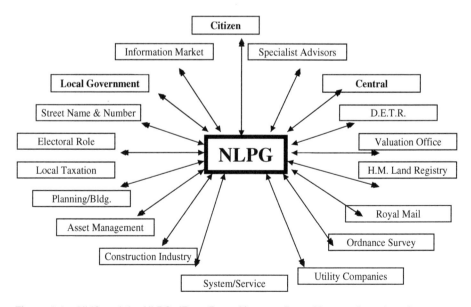

Figure 9.1   NLIS and the NLPG. (From Bruce Yeoman, Bruce Yeoman Associates.)

At this point it is worth noting that in 2000, TerraQuest undertook a reappraisal of how the anticipated costs and benefits contained in the original business case had translated into reality. This showed that the targeted net positive cash accumulation had been achieved, and exceeded, through the additional fees income that had accrued to the trading account due to the increase of searches that had occurred — up to 12,000 from 1999 to 2000, compared to 9,000 searches that had previously been assumed. These searches had been dealt with by the reduced level of staff as envisaged in the business case without detriment to achieving the anticipated reduction in search turnaround times.

Stage 3 (1999 to 2002) — Following the success of the NLIS pilot, this has been a time of consolidation for the council. Because of their direct experience, officers of the council remain heavily engaged, together with the IDeA, in supporting the promotion of NLIS. The council has continued to be involved in closely related projects and has taken part in the DETR's trials of the proposed House Buyers/Sellers Pack. The council has also begun an important project, in support of NLIS, to voluntarily register its property ownerships with the HMLR (25 to 40% of the city is owned by the council). Once this project has been completed, anyone who wants to acquire land or property from the city council will simply be able to obtain a copy of the title documents from the HMLR rather than having to go through the current long drawn-out process of having to plough through a time sequence of related deeds in order to prove title.

In addition, the council has deliberately broadened the focus on GIS away from LLC to the other service areas that can potentially benefit. Internet-based GIS and associated access to land and property data have been rolled out to all departments based on Bristol City Council's LPGWeb (which is an Internet-enabled gazetteer linked to OS maps and a high-interest subset of the council's spatial data: see Figure 9.2). A metadata database (register of spatial datasets and projects) has been compiled through an extensive information audit of all departments, which is backed up by an electronic library of GIM standards and how to use advice as prescribed by the council.

9.3 WHAT ORGANIZATION HAS IT SET UP?

In 1993 (following an organizational development study within the council, which was undertaken by INLOGOV) the first forum for steering and coordinating corporate working with regard to land and property information was began by organizing the property and land user group (PLUG) (later renamed the Land and Property Working Group) which contained second-tier representation from each of the major departments. This group was instrumental in setting the highest priority for the implementation of a central land and property referencing system within the council.

After local government reorganization in 1996, the group was replaced by the geographic information group (GIG), again at second tier. It was responsible for setting the overall strategy for, and coordinating the management of, land and property information within the council and monitoring (at a high level) the implementation of specific projects (such as LLC and corporate GIS). The group was

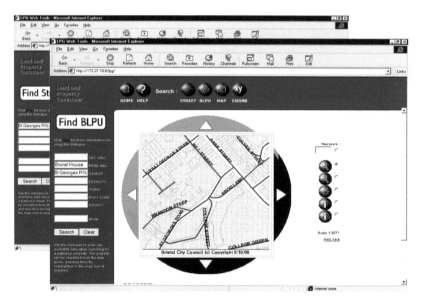

Figure 9.2   (See Color Figure 1 following page 134.) Bristol City Council's Web-based approach that allows the rapid location of properties of interest by all staff across its intranet. (From Bristol City Council. Reproduced with permission from Ordnance Survey. © Crown Copyright NC/03/16653.)

supported by a separate corporate property group that was responsible for putting in place the framework for the more detailed day-to-day management of information relating to corporate property interests and assets.

Despite the high profile for GIM within the authority, there is no central unit that supports departments in the use of GIS and spatial information. Each department has made provision, as needed for its own resident GIS skills, drawing where appropriate upon the skills of neighborhood and housing services, which has emerged as a "center of excellence" in ESRI GIS products (on a charged-for consultancy basis).

9.4 WHAT DOES BRISTOL CITY COUNCIL PLAN TO DO IN THE FUTURE?

For the future, the council is focusing upon the provision of easily accessible, low-cost, maintainable GIS for use throughout all departments. The LLPG is currently accessible online by all departments across the council's intranet. The city council is now in the process of upgrading its LLC software to the new Total Land Charges (TLC) module supplied by ESRI, and at the same time upgrading its CAPS uniform software to Spatial Version 7. The project will bring into one GIS suite an even wider range of services that previously used different GIS products or had no GIS facilities at all. This will greatly enhance the degree of interoperability of systems and the sharing of data. The system as a whole will continue to be underpinned by the corporate LPG, which will now also be delivered through the appropriate uniform module (GMS).

The future strategy for GIS in the city council is being reviewed in the context of an interdepartmental structure that has now stabilized after several years of reorganizations. It is likely to focus on using CAPS uniform systems as the key vehicle for the further development of its public interfaces and for meeting the city council's e-government targets.

9.5 WHAT WERE THE POSITIVE DRIVERS AND SUCCESS FACTORS FOR GIS AND GIM?

Bristol City Council has enthusiastically taken the lead on a GIM project of high visibility and national significance. It has been opportunistic in taking advantage of an alliance of like-minded external interests to achieve an information infrastructure that has dramatically improved the way it has been able to deliver land and property based services. The most important positive drivers that have encouraged the council's pioneering efforts in GIM have been:

- The climate of cut-backs in services in the early and mid-1990s that forced the council to be imaginative in looking at how improvements in land and property information could be used to support enhanced service delivery
- The growing needs within the engineering function for comprehensive street and property data, and within the planning function for computerized map-based information
- The increasing recognition of the importance of creating an authoritative corporate LPG as the means to facilitate an unambiguous exchange of information about a property or plot of land within the city
- The coincident timing of the publication of BS7666 as the standard for LPGs
- The parallel invitation from LGMB to join the NLIS pilot, which provided access to external skills and resources
- The push to action from individual champions within the city council that spurred on the efforts of staff

In addition, a number of critical success factors, have reinforced the above drivers, to make the project a success:

- Positive interpersonal relationships among all involved in the NLIS project that contributed to the establishment of a cohesive team
- Benevolent senior management within the council who adopted a hands-off approach and delegated the key decision making to technically skilled staff
- High levels of skill and an awareness of data management issues within the core staff that were major advantages in getting new ideas off the ground

9.6 WHAT WERE THE NEGATIVE FACTORS THAT THREATENED SUCCESS?

Despite the long list of drivers and success factors for the project, there were a number of negative factors that potentially threatened success:

- The constant undercurrent of reorganization, which introduced an air of change and uncertainty across the council
- Lack of commitment of resources to GIS from the top level, despite an awareness of the potential benefits
- Preoccupation of middle management in satisfying short-term information needs at the expense of longer-term investment in the infrastructure

9.7 WHAT HAVE BEEN THE PRACTICAL BENEFITS?

GIS and improved GIM in Bristol City Council has delivered a wide range of benefits:

- Computerization of LLCs has been achieved on a self-financing basis (see Section 9.2) and has resulted in response times being more than halved (to 5 days immediately following computerization, and now down to 3 days per search) despite a 50% reduction in staff. This happened over a period when volumes rose by 55% from 9,000 searches per annum prior to computerization to 14,000 searches per annum.
- The LLPG is now in wide use throughout the council, enabling the exchange of information about properties and plots of land on a common basis. The LPGWeb that operates over the council's intranet is heavily used by all departments to identify streets, properties, and plots of land and obtain associated information.
- The wider benefits from the publication of data for use by different systems is yet to be realized, but early experience of the Lighting Web System has resulted in reduced duplication of effort, increase in officer time through efficient access to information, and improved quality of information for decision making.
- The metadatabase is improving efficiency and effectiveness by reducing the time required to locate and obtain information.
- Systems designed for use by members of the public (see Faulty Light Reporting and Abandoned Vehicles Reporting at www.bristol-city.gov.uk) make practical use of the LLPG, and enable structured information to be rapidly entered.
- GIS and the LLPG are providing improved communication channels, enabling quick and efficient access to information both within the council and in its dealings with the public and external agencies.
- Participation in the NLIS pilot has ensured that the council is seen as dynamic and go-ahead and has been widely publicized as such but in a way that is firmly rooted in achieving practical improvements in services for its own residents.

9.8 WHAT ARE THE LESSONS FOR OTHERS?

Most authorities shrink from being at the cutting edge of technological developments, but they would find it extremely difficult to move forward without being able to follow local authorities such as Bristol City Council that have relished the pioneering role. As the local authority that has been the focus of the NLIS pilot, this case study of Bristol City Council has a number of very important messages for others:

- For those few local authorities that are keen to become pioneers, the case study makes clear how important it is to ensure that there is a sufficiently strong and burning need to undertake developments that currently do not exist or have not been fully tested in order to make the added risk worthwhile. When this is so, experimental projects can often provide the route to enlisting external resources that would not normally be available and making things happen in a shorter time.
- For those many local authorities that prefer to watch and wait, the NLIS pilot study is well documented (see www.nlis.org.uk) and the NLIS and NLPG programs that are currently being rolled out have incorporated many of the lessons learned through the experiences of Bristol City Council. The key messages to such authorities are:
  - To consider and review the call to participate in the NLIS and the NLPG programs but to follow Bristol City Council's example and do so, first and foremost, in a way that ensures that local benefits are rapidly obtained from the improved ability to compile and exchange land and property information.
  - To derive the highest possible benefits from participating in NLIS and NLPG versus the costs by fully capitalizing, as Bristol City Council has done, upon the land and property data that has been expensively collected (and the supporting infrastructure that has been put in place) by ensuring that it is widely accessible across the local authority and from outside.

CHAPTER 10

Case Study — Southampton City Council

SOUTHAMPTON CITY COUNCIL AT A GLANCE

Key Facts

Local authority name: Southampton City Council

Local authority type: Unitary council

Population: 216,000

Current state of operation of GIS: Multi-supplier/Authority-wide GIS

Main GIS products in use: Intergraph Geomedia (10 data management/edit licenses plus Web browser access), which has recently been adopted as the corporate GIS. MapInfo Professional (98 licenses).

Applications: Land charges, building control, development control, street lighting, environmental health, trading standards, city safety/crime reduction, school catchment area planning and "safe routes to schools," child care, housing stock, contaminated land, and land terrier

Land and Property Gazetteer status: CAPS uniform gazetteer that is BS7666 compliant

GIM/GIS strategy status: No formal GIS strategy

Forum for steering GIS: No GIS steering group but a "GIS User Movement" meets occasionally for briefings on current projects and new developments

Staffing for GIS: IT section provides implementation and ongoing support and training

Contact details: GIS business analyst (telephone 023 808 33092)

What Makes Southampton City Council Distinctive?

Southampton City Council is an admirable example of an authority that, typical of many, has adopted a grassroots approach to GIS. Its approach to implementing GIS has been ad-hoc and step-by-step, and has been led by user activity on specific projects of interest, which has taken place without the framework of a council-wide GIS strategy and without the direction and coordination of any GIS steering group (but with the support of users who banded together to form the Underground GIS User Movement). Despite the initial lack of any corporate approach, the council has been successful in the application of GIS, has a number of showcase projects, and the case study holds many lessons for those local authorities whose style and culture has encouraged departments to go it alone without the luxury of diverting time onto overall planning and coordination on an interdepartmental basis. Recently, a corporate approach has begun to emerge with the adoption of Intergraph Geomedia as the corporate GIS.

Key Stages in the Implementation of GIS

Southampton City Council's experience of GIS has taken place over three main phases:

Stage 1 (1980s to early 1990s) — Piloting and rejection of IBM's mainframe GFIS in the 1980s, followed by initial interest, then waning commitment to Intergraph's Microstation in the early 1990s

Stage 2 (1997 to 2000) — "Organic" emergence of GIS applications (based on MapInfo) that had sprung up independently "like a field of mushrooms" since the authority became a unitary

Stage 3 (2000+) — Internet-enabling of applications and recent emergence of the corporate approach with choice of Intergraph Geomedia as the corporate GIS

Positive Drivers and Success Factors for GIS

- Individual officers in departments identified projects where GIS could provide significant benefits and implemented them without the red tape of large corporate projects and at low cost on desktop PCs.
- In-house IT section (in computer and printing services department) provided a packaged GIS service at only £1,500 "one-off" cost per desktop PC (covering MapInfo software, training, GIS support, OS map-base/updates, and access to other department's spatial data).
- Widespread acknowledgment of success of GIS in education.

Problems that Threatened Success

- GIS has had to flourish through an "underground movement" as a consequence of the lack of senior management and corporate support.

Practical Benefits from GIS

- Availability of one up-to-date corporate map base and sharing of data across departments.
- Self-contained benefits within each project in the form of improved efficiency and quality of decisions through the ability to exploit spatial aspects of data relating to clients and services.
- Education became an outstanding example of what could be achieved at low cost. Use of GIS revealed the "looser" real-world catchments caused by exercising of parental choice and encouraged data sharing and the avoidance of duplication of effort that resulted from collaboration with the transportation division.

10.1 WHY WAS SOUTHAMPTON CITY COUNCIL CHOSEN AS A CASE STUDY?

Southampton City Council was selected as a case study because it is typical of many local authorities. Despite a long history of experimentation with GIS, which started in the 1980s with piloting and rejection of IBM's mainframe GFIS GIS and extended in the early 1990s into experimentation followed by waning interest in Intergraph's Microstation, it is only since Southampton's establishment as a unitary authority in 1997 that GIS has really taken off. Faced with the fact that its early attempts at using GIS had failed to establish a corporate approach, its more recent experience in GIS has deliberately been ad-hoc and step-by-step, and has been led by user activity on projects of interest that have sprung up "organically" in departments, e.g., the use of GIS in education for catchment area planning and safe routes for schools, which has been a showcase project to the rest of the authority. These user-led projects have shown considerable success despite the absence of a council-wide GIS strategy or the direction and coordination of any GIS steering committee. Although a corporate GIS steering committee was briefly established on local government reorganization in 1997, it folded through lack of support after only one meeting, to be replaced by the emergence of the GIS Underground User Movement that was established by the users as a basis for sharing experiences and discussing common issues and has been a key mechanism for getting GIS off the ground. Only recently has the council begun to regain its corporate approach with the widespread use of Intergraph Geomedia as the corporate GIS.

The case study holds many lessons for those local authorities whose style and culture have encouraged departments to go it alone without the luxury of diverting time into overall planning and coordination on an interdepartmental basis. The case study shows the considerable success with GIS that can be achieved without a strong corporate approach. However, it also recognizes that the sharing of spatial data between departments can take place effectively only within the framework of corporate standards. This has resulted in GIS recently moving full circle with the emergence of a corporate approach to GIS within the council based upon the use of Intergraph Geomedia.

10.2 THE BACKGROUND — WHAT HAS SOUTHAMPTON CITY COUNCIL DONE?

Southampton City Council is an example of the implementation of a multi-supplier/authority-wide GIS, using the terminology that we introduced in Chapter 8. Its experience of GIS since 1997 has focused initially on the introduction of software from MapInfo Corporation's GIS portfolio, but recently Intergraph Geomedia has emerged as the corporate GIS.

Currently, the following GIS software products are in use across the authority:

- **MapInfo Professional** (98 licenses) is used across the authority, but note that although all data is accessed from a central GIS Novell server, it is not a corporate system as all licenses were purchased in ones and twos as required by individual projects. With the decision to use Intergraph Geomedia as the corporate GIS, no additional MapInfo licenses are being procured.

- **MapInfo MapXtreme Web server** enabling more than 2000 users to access GIS via the intranet.
- **Integraph Geomedia** (10 data management/edit licenses with widespread browser access throughout the council) was installed in the early 1990s for initial use for grounds maintenance contract and only recently has been selected as the corporate GIS.

Within Southampton City Council, development and implementation of GIS has taken place over three major stages:

Stage 1 (1980s to early 1990s) — This was a period of early experimentation with GIS. Starting with a brief flirtation with IBM's mainframe GFIS GIS in the 1980s, which was piloted and rejected as unsuitable for corporate use, the council at the end of the 1980s and early 1990s committed itself to an Intergraph Microstation system. This was tested initially within the leisure services department for the preparation of grounds maintenance invitations-to-tender (for which the production of maps showing the boundaries of grounds had previously consumed much manual effort). Selected with the potential to become the corporate GIS, significant budgets were spent on Unix servers and workstations, training, and development, including the installation of further seats within the highways and transportation division.

Rich in CAD, 3-D, and solid modeling functionality, the Intergraph Microstation proved to be a powerful tool in the hands of expert users, but overkill for typical local government users. According to Adnitt (1998), It was "over-complicated and difficult to use, the simplest tasks were awkward and cumbersome, the situation further aggravated by the fully trained expert user leaving to seek pastures new. Frustrated users, desperate for maps for operational support, resorted back to paper and coloured pencils, leaving the workstations to gather dust." The result was the demise of the Intergraph Microstation, which continued to be used only within the leisure services department for the infrequent preparation of grounds maintenance contracts.

Stage 2 (1997 to 2000) — With the creation of the unitary authority in 1997, GIS entered a new beginning. The cost of the OS copyright licenses under the Service Level Agreement (SLA) rose dramatically from £12,000 to £32,000 as a result of the increased size and responsibilities of the authority. At the outset, it was recognized as critical that as many potential users as possible were recruited to GIS, and a corporate GIS steering committee was established to encourage takeup and ensure proper control and coordination. Disappointingly, this group never met again after its inaugural meeting, with senior management representatives becoming preoccupied with reorganizing their divisions to cope with the new pressures upon them coupled with "a natural skepticism for all things corporate, especially IT!" The views of many at that time can be typified in the comment that "if there is a sure way to kill a project stone dead it is to give it the corporate label. For corporate read big, expensive, complex, too late to meet the business need and not what was requested in the first place. By the time the analysts have analysed, project managers managed and the methodology controlled, the business has moved on and the solution is designed for a problem which has changed or no longer exists" (Adnitt, 1998).

It became clear that progress toward a corporate GIS was some way off, and in order to avoid total loss of momentum, the computing and printing section introduced a standard packaged service for the implementation of GIS based on MapInfo and covering for a one-off cost of £1,500 for each desktop PC:

- MapInfo software, procurement and updates
- Training
- Support and consultancy
- OS map distribution and updates
- Access to corporate map server
- Access to other departmental spatial data
- CD-ROM distribution for stand-alone PCs
- Services of the council's Ordnance Survey liaison officer (OSLO)

This "bargain offer" from an enlightened IT section has greatly encouraged the use of GIS, and with a single set of accurate up-to-date map data on the corporate network server, over 45 desktop PCs had signed up by the end of 1999 for the following applications that emerged "organically" in departments like a field of mushrooms:

- **Education and transportation** — for schools catchment area planning and safe routes to schools initiative
- **Highways and street lighting** – all 27,000 lighting columns, lights, illuminated signs, and bollards surveyed with data held in MapInfo interfaced to its Confirm (highways and street lighting) system from Southbank Systems
- **Public sewer records** — supplied on CD from Southern Water and converted to MapInfo
- **Child care strategy** — comparing service provision geographically with need
- **Planning, building control, trading standards, and environmental health** — implementing portfolio of CAPS uniform systems linked to MapInfo
- **Housing** — plotting locations of council housing stock in response to new Crime and Disorder Act

Much of what stimulated the emergence of this list of GIS applications was the early selection of a showcase project that could be used as a demonstrator to other departments. The identification, implementation, and promotion of education as the initial "killer" application for GIS have underpinned the widespread expansion of GIS across departments. In 1997, the education department was looking to define a 5-year strategic plan for the provision of education, an important basis for which was the location of schools and school catchment area boundaries, for which a critical need was to know where pupils lived in relation to the schools they were attending. Every school was asked to supply the education department with a floppy disk containing the address of every pupil in its SIMS (school records) system. These were then grid-referenced by the IT section using OS ADDRESS-POINT data in order to create maps in MapInfo for the existing catchment areas. These maps immediately confirmed the effects of parents exercising their rights to choose their children's schools by revealing the extent to which previous "tight" catchments had frequently (particularly in the case of the more desirable schools) been replaced by "looser" catchments with longer average journeys to school. The data were also further processed by the transportation division to support a "safe routes to schools" initiative for the 27,000 journeys that started at about the same time by car, bus, train, bicycle, or foot. The above uses of GIS in education demonstrated to other departments what was achievable in terms of improved insight from the mapping of data in relation to services, and the benefits that accrued from sharing of data and avoiding duplication of effort through collaboration with the transportation division.

To fill the vacuum left by the disbanding of the corporate GIS steering committee and to provide a mechanism to share and learn from one another's experience, the users banded together to form the Underground GIS User Movement that played a key role in encouraging the expansion of GIS. Its explicitly anti-establishment and contentious title ensured its popularity among day-to-day users as a forum that met every 2 or 3 months to allow the airing of views. A measure of its success is that by the end of 1999 it had dropped the word *underground* from its title. As a further means of disseminating information, the movement produced the regular *GIS News-letter* that was:

- Widely distributed across the authority
- Written in a light, easy-to-read, and nontechnical style
- Visually interesting with the inclusion of graphics and cartoons
- Packed with news about latest developments on projects in departments (which undoubtedly had the effect of encouraging nonactive departments to join the GIS club to avoid being left out)
- Used to promote forthcoming meetings of the GIS User Movement and to stimulate attendance of new members

Stage 3 (2000 to present) — The last stage in the development and implementation of GIS in Southampton City Council has been characterized by the e-enabling of GIS applications and the reappraisal of the need for a corporate approach. By 2000 there were 98 MapInfo licenses installed across council departments, allowing access to spatial data on the Novell server across the council's network. The many applications installed in Stage 2 had been further extended by use of MapInfo for:

- **Contaminated land** — establishment of map-based register to meet the requirements of Section 57 of the Environment Act 1995
- **Land terrier** — digitization of the council's property portfolio from paper maps
- **Single Regeneration Budget (SRB)** — digitization of SRB areas to provide mailshot addresses of people on benefits
- **Winter salting routes** — digitizing of routes treated in 2000 in order to identify any key roads omitted or duplicated and any conflicts in routes

Access to the rich pool of data available from Southampton City Council's GIS applications has been greatly enhanced through use of Internet and intranet technology. In April 2000, the council invested in MapXtreme (MapInfo's Internet map server), hosted on a dedicated MS Windows NT server, to provide access to most

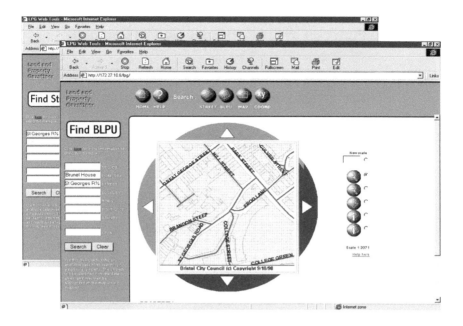

COLOR FIGURE 1 Bristol City Council's Web-based approach that allows the rapid location of properties of interest by all staff across its intranet. (From Bristol City Council. Reproduced with permission from Ordnance Survey. © Crown Copyright NC/03/16653.)

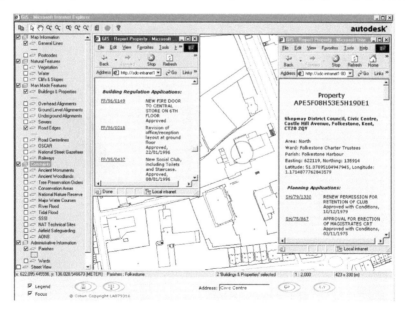

COLOR FIGURE 2 Shepway District Council's planning database showing planning and building regulations applications at the Civic Centre, Folkestone. (From Shepway District Council. Reproduced with permission from Ordnance Survey. © Crown Copyright NC/03/16653.)

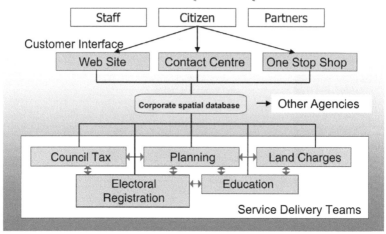

COLOR FIGURE 3 London Borough of Enfield's vision for the delivery of data from service information systems to the council's customers in conjunction with its partners. (From London Borough of Enfield.)

COLOR FIGURE 4 New technologies: GIS on handheld devices delivering data and services.

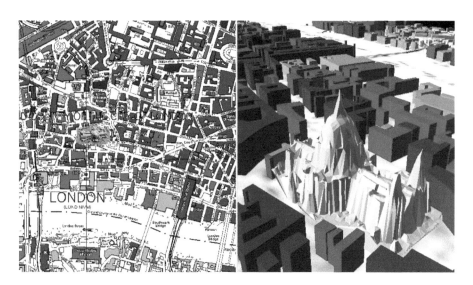

COLOR FIGURE 5 The way we might visualize and navigate through digital reconstructions of real cities: An example in St. Paul's district of the city of London, using light imaging (LiDAR) data in 3-D GIS. (Reproduced with permission from Ordnance Survey. © Crown Copyright NC/03/16653.)

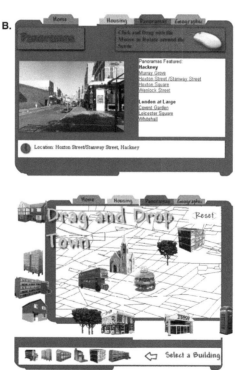

COLOR FIGURE 6 (Environmental and educational geographic information systems: (**A**) Querying pollution information at specific sites within London. (Reproduced with permission from Ordnance Survey. © Crown Copyright NC/03/16653.); (**B**) A Web page from the Hackney Building Exploratory Interactive System for educating the public about their local environment.

of the mainstream GIS applications through a Web browser across the council's intranet. This now supports access to GIS from over 2000 staff within the authority, but has not yet been opened to the public.

Although Southampton City Council describes its approach to GIS as grassroots rather than corporate, it is important to acknowledge that the IT section (within computing and printing services) and the GIS User Movement have helped to put in place a framework for the coordination of procurement, project implementation, and data sharing, which though low key, has achieved many of the aims of a formal corporate approach. The wheel has turned full circle and Southampton City Council has now begun to reappraise the need for a corporate approach to GIS, of which the first step has been the recent selection of Intergraph Geomedia as the corporate GIS. This has 10 concurrent licenses (for data management and editing), but the main users now have access to maps through a Web browser interface over the corporate network or through the use of GeoMedia Objects embedded in council-developed applications (for bin collection, trees, and grounds maintenance).

10.3 WHAT ORGANIZATION HAS IT SET UP?

As described above, Southampton City Council set up a corporate GIS steering committee on reorganization in 1997, with membership from senior managers. However, it only met once because attention was diverted to the practical problems of reorganizing the authority to cope with an extended range of services and because of a distrust of all things corporate.

The continuing pressure from users and potential users for a forum for discussion led to the creation of the Underground GIS User Movement (with the word *underground* eventually dropped from its title), which met to exchange news, experience, and views. The movement has had a very important role in introducing an informal authority-wide framework within which individual GIS projects have been implemented, and has, in effect, achieved an undercover corporate approach.

The IT section (within computing and printing services) has also been a very important factor in promoting the takeup of GIS within the authority in an enlightened manner. Its two staff allocated to GIS (GIS business analyst and analyst programmer) have taken a lead role in establishing the GIS User Movement, preparing and issuing the *GIS Newsletter,* and stimulating GIS projects through the provision of a low-cost packaged GIS service for desktop PCs.

10.4 WHAT DOES SOUTHAMPTON CITY COUNCIL PLAN TO DO IN THE FUTURE?

In the future, Southampton City Council intends that all GIS users, no matter what application or interface they are using, will access a single GIS data repository that will reside in Oracle. The preferred GIS for corporate use will be Intergraph Geomedia, which far surpasses the earlier Intergraph Microstation GIS, and will be the product on which most GIS development in the future will be based.

The existing MapInfo users will have direct access to the GIS database, thus protecting Southampton City Council's past investment (licenses, training, and developed applications) without duplicating the data for use by different GIS applications.

10. 5 WHAT WERE THE POSITIVE DRIVERS AND SUCCESS FACTORS FOR GIS?

Despite not having a strong corporate approach, Southampton City Council has been very successful in introducing a wide range of GIS applications across all major departments in a short space of time. Almost as a positive reaction against its failed attempt to introduce a GIS steering committee, a number of drivers have stimulated investment in GIS, in particular:

- The emergence of the slightly shady sounding Underground GIS User Movement that captured the imagination, and, more importantly, the support of users and potential users for sharing experience and information about GIS
- The resulting groundswell of users who put their energies into implementing a wide range of GIS projects that fit within their budgets
- The self-evident success of the education project as a showcase of what GIS could potentially achieve

One of the critical success factors fundamental in underpinning the implementation of GIS has been:

- The leading role played by the IT section in promoting the GIS User Movement and in encouraging takeup of GIS through the offer of an attractive, low-cost GIS support package (which negated the cost-benefit arguments prevalent in many local authorities)

10.6 WHAT WERE THE NEGATIVE FACTORS THAT THREATENED SUCCESS?

The most important serious negative factor that has threatened the success of GIS from the outset has been lack of senior management and corporate support. However, as Section 10.5 has suggested, this has become a positive stimulus to building an undercover movement that has done much to get GIS underway.

10.7 WHAT HAVE BEEN THE PRACTICAL BENEFITS?

Although no comprehensive and systematic audit has been undertaken of what has been achieved from the many implemented GIS projects, it is clear that GIS has delivered a wide range of benefits:

- Availability of one up-to-date corporate map base across the authority with savings of resources in departments that previously maintained and updated their own map sets to differing levels of consistency
- Acceptance that the sharing of data across departments was to the advantage of all and provided a basis for joint collaboration on issues of common concern and the avoidance of duplicated effort
- Recognition of education as an outstanding example typifying the benefits that could be achieved from improved ability to exploit the spatial dimension of data relating to customers, services, and transport infrastructure

10.8 WHAT ARE THE LESSONS FOR OTHERS?

This case study holds a number of lessons for local authorities that, like Southampton City Council, have found it difficult to put in place a corporate approach to the implementation of GIS. It emphasizes the importance of:

- Using a reaction against the corporate approach as a positive stimulus to ensure that a less formal working-level forum for discussion on practical issues relating to GIS is installed
- Diverting collective energy that will no longer be consumed by (often protracted) corporate discussions into making things happen at the departmental level in terms of GIS projects that directly contribute to service priorities
- Making sure that, even with the absence of a corporate approach, the economies of large organization purchasing are still achieved and imaginative ways to promote the start-up of GIS are encouraged (e.g., the IT section's prepackaged low-cost GIS implementation service)
- Carefully selecting and promoting a showcase GIS project (e.g., education catchment analysis and safe routes to schools) that demonstrates to other departments what can be achieved

As a final comment, although the case study shows the considerable success with GIS that can be achieved without a strong corporate approach, it also highlights the importance of corporate standards in ensuring an effective framework for the sharing of spatial data across departments. Southampton City Council has already begun to reintroduce its corporate approach to GIS based upon Intergraph Geomedia.

CHAPTER 11

Case Study — Leeds City Council

LEEDS CITY COUNCIL AT A GLANCE

Key Facts

Local authority name: Leeds City Council

Local authority type: Metropolitan borough

Population: 715,400

Current state of operation of GIS: Multi-supplier/Authority-wide GIS (but note that most GIS implementations are confined to specific departments with few inter-linked corporate applications)

Main GIS products in use: ESRI's ArcInfo 7, ArcView and MapExplorer2 (with ArcGIS and ArcGMS under test); MapInfo; and GGP

Applications: Map production, land terrier, LLC, planning and building regulations application processing, and highways design (plus a variety of small stand-alone systems)

Land and Property Gazetteer status: Largely conformant to BS7666 (populated from the Local Authority Management Information System (LAMIS) Gazetteer)

GIM/GIS strategy status: Agreed to using one supplier to implement a central corporate gazetteer that will be available to all departments and to enable public Internet access to some data sets

Forum for steering GIS: GIS program board with higher level business, IT, and strategy group that prioritizes funding and resource allocation

Staffing for GIS: GIS program manager (in IT services)

Contact details: Principal planning officer (data) (telephone 0113 247 8122)

What Makes Leeds City Council Distinctive?

In the early 1970s, Leeds Corporation (as it was called then) signed up as the pilot site to develop and implement ICL's LAMIS. Starting from an early desire to introduce an overall corporate approach to land and property information, the council then entered the doldrums of disillusionment before reconceiving its approach to gazetteers and GIS from the 1990s onward.

Key Stages in the Implementation of GIS

Stage 1 (1972 to 1980) — Pioneering LAMIS project (construction of UPRN hub file) that was a joint venture among Leeds City Council, ICL, and the Department of Trade and Industry

Stage 2 (1980 to 1991) — LAMIS project not completed and key users such as rates disconnect from the hub. Planning and building regulations applications processing and housing maintenance continue to use LAMIS

Stage 3 (1991 to 1996) — Maintenance of property hub file transferred to planning department. Corporate GIS project assesses the potential for GIS throughout the council. As a result, GIS-based systems for land terrier and local land charges systems installed (but local land charges system not live until 2003 because of system performance and data quality problems)

Stage 4 (1996 to 2002) — Desktop revolution encouraged many departments to purchase GIS for individual applications. No strict single supplier policy but most systems based on ArcView. In 2001 Leeds City Council committed to adopting a single supplier and to setting up a corporate gazetteer

Positive Drivers and Success Factors for GIS

- Early participation in LAMIS and associated opportunities for cost sharing
- Severe deterioration of paper base maps for local land charges register and land terrier
- Commitment of high-level champions such as director of finance
- Reduced computer storage costs and higher processing speeds that lowered entry level to GIS
- Increased pressure for access to information coupled with availability of easy-to-use GIS software
- Availability of digital OS base through the SLA (from 1994)

Problems that Threatened Success

- Lack of sustained funding for LAMIS arising from underestimation of costs and technical challenges
- Extended timescale for implementing LAMIS with lack of results and loss of credibility
- Over-run of data capture projects
- Lack of awareness of the potential of, and opportunities for, GIS among senior managers

Practical Benefits from GIS

- Improved quality of data through more extensive usage
- Graphical power of visualizing topical data within GIS
- Map location now the "common currency" for exchanging data across departments

11.1 WHY WAS LEEDS CITY COUNCIL CHOSEN
AS A CASE STUDY?

In the early 1970s, Leeds Corporation (as it was then called) committed to an ambitious project to form a LAMIS based on the principle of creating a number of banks of information that would be shared across the different local authority functions. The vision was to establish a comprehensive local authority database that was management oriented and kept up-to-date by the day-to-day operational tasks of the authority. It was recognized that this all-encompassing vision would take several years to deliver benefits, and the authority took an "act of faith" to pilot the development of the system without the anticipation of significant early returns.

For the first phase of the project, which involved information relating to land and property, the Leeds Corporation invited ICL to join it to form a joint team. Together with the resources of the corporation, funding was obtained from the Department of Trade and Industry and ICL. The aim of the project was to develop a solution that was applicable as far as possible to other authorities and to report on the problems likely to be encountered in implementation.

This pioneering attempt to introduce an overall corporate approach to information systems started bravely and enthusiastically in 1972 with the undertaking of a 3-month review of all council functions that could potentially benefit from the improved availability of information. On the basis of the resulting report, the council decided to put in place as a first priority a common database supporting the rating, planning, local land charges, and housing rents functions with, at its core, the establishment of a hub file containing all the properties within Leeds, each identified by a UPRN. But by the early 1980s the project had still not been completed, and disillusionment set in as a consequence of widespread recognition that it was failing to deliver. Part of the reason for this failure lies in the penalties that Leeds City Council paid for being an early pioneer. In particular:

- Spatial data was limited in availability. At the time LAMIS was under development in the early 1970s, there were no digital maps, so spatial data had to be digitized from paper maps, and then plotted for checking as overlays that could be super-imposed on the paper maps.
- Computers were in their infancy, so computer storage capacity and associated costs were major problems.
- Spatial analysis techniques (now taken for granted as a part of GIS) were only just being developed, and the early GIS specialists were still grappling with problems such as how to write script that could tell whether a point was inside or outside a polygon.

Leeds City Council is distinctive because it is a prime example of an authority that began a grand design in information systems in the 1970s only to recognize too late that projects that rely on lengthy timescales for delivery, without the careful selection of "early winners," are doomed to failure at the start. But Leeds City Council, together with others such as Bradford and Coventry, did much to demonstrate the potential value and identify the practical problems of establishing land and property gazetteers. Their groundwork has been subsumed into the approaches of other local authorities and the emergence of BS7666. To have come so far in developing a complex data management system such as LAMIS, while working within the severe constraints identified above, needs to be recognized as an outstanding achievement despite the project's lack of completion.

Following retrenchment into departmentalism in the 1980s and early 1990s the phoenix has again begun to reemerge from the ashes. From the mid-1990s the successor Leeds City Council has begun to build upon its past investment and knowledge and move gently toward resurrecting the infrastructure that could eventually support a corporate approach.

11.2 THE BACKGROUND — WHAT HAS LEEDS CITY COUNCIL DONE?

Leeds City Council is an example of the implementation of a multi-supplier/authority-wide GIS, using the terminology that we introduced in Chapter 8. It currently has a mix of software from ESRI (ArcInfo 7; ArcView, and MapExplorer2), MapInfo Corporation (MapInfo Professional), and GGP across the council. However, while there is widespread use of GIS, it is important to note that most GIS implementations are confined to specific departments (or even parts of departments) with few interlinked corporate applications.

The major databases and processing systems that are linked to GIS include the LPG, land terrier (covering the city council's land holdings and transactions), LLC register, and planning application processing system. These systems have taken several years to become operational. The planning and environment department is already in the course of procuring a replacement planning application processing system in order to keep up with rising user expectations of what a modern computer system must be able to offer.

The LPG is live and being maintained, but the bulk of the data was loaded from the former LAMIS UPRN hub file and needs substantial cleaning up. Several important flows of information on changes, in particular council tax and Non-Domestic Rates, are not yet interfaced to the gazetteer. The gazetteer is largely conformant with the first (1994) version of BS7666, but has not been revised to take into account the changes required by BS7666 (2000). The gazetteer has been implemented as part of Phase 1 of the corporate GIS, which is based heavily on ESRI's products, and is an integral part of the LLC application. Access to the gazetteer by other view-only users is restricted by the technical difficulties involved in sharing the data that is currently locked into the local area network that supports LLC and the land terrier. A prototype BS7666 (2000) conformant gazetteer — based on the Phase 1 original gazetteer extended by data from the housing stock information base, council tax and National Non-Domestic Rates, and electoral register — has recently been established.

In addition, all the land terrier data have been loaded and the system is now live. The LLC data were loaded in December 2000–January 2001 with live running now imminent. The planning application processing system is free standing but can be accessed (in "read only" mode) by the LLC register.

Looking back at the history of the development and implementation of GIS and improved access to spatial data in Leeds, it is clear that implementation has taken place over four major stages:

Stage 1 (1972 to 1980) — The LAMIS project started as a joint venture among Leeds Corporation, ICL, and the Department of Trade and Industry, with the close involvement of the finance, planning, housing, and highways departments. The project involved the construction of a hub file of the city's properties (known as Basic Spatial Units or BSUs), each of which was identified by a UPRN and an associated grid reference, and which was intended initially to support the business systems for rating, planning, local land charges, and housing rents. The UPRN was held in the format:

$$SSSSS/PPPP/I/XXX$$

where S was the street code, P a property within the street, I a pseudo-indicator to indicate that a property was not identified by a number (e.g., identified by a house name held on a separate names file), and X a sub-subdivision (for a property split into parts, e.g., flats).

The methodology for constructing the hub file was based heavily on parallel work led by the Department of the Environment to develop a manual on point referencing properties and parcels of land (for which the draft was available at the start of the LAMIS project in 1972).

While the hub file of over 200,000 BSUs was established by mid-1973, including the digitizing of BLPU centroids, the work required to create the links to the business systems (rating, planning, local land charges, and housing rents) became protracted and fraught with difficulties. The wide-reaching nature of the project made ongoing corporate commitment and continued input of the necessary resources difficult to sustain as a multi-year program in the face of other pressures on the local authority (e.g., local government reorganization in 1974). By the end of the 1970s the project was only partially completed, and support evaporated with the lack of significant tangible benefits. At this time some links to rating, planning, and housing systems had been achieved, but the local land charges function remained unsupported.

Stage 2 (1980 to 1991) — During the 1980s the council entered the "quiet years" in terms of a disenchantment with corporate initiatives relating to land and property data. Further development of LAMIS in Leeds was halted in 1981 (as a consequence of the council's upgrading its mainframe from ICL to Honeywell), but ICL continued to work with a small group of local authorities elsewhere, e.g., in Dudley and Thameside. As a consequence, the rates department decided to go its own way by disconnecting from the hub file. Despite the investment that had been made in establishing a potentially corporate infrastructure for land and property data in the form of the hub file, the only uses of the LAMIS system were for some individual operational purposes. In particular, the main uses were for planning and building regulations applications processing and some specialized housing functions such as painting and maintenance scheduling. Over this period, the LAMIS hub file continued to be maintained for these users by the IT section but as a low priority. A gradual skills decay occurred as IT staff who were familiar with the system left and were not replaced, making it extremely difficult to modify the system to take into account new legislative and operational requirements.

Stage 3 (1991 to 1996) — The first half of the 1990s was characterized by a resurgence of interest in land and property data, forced by a growing enthusiasm for exploring what GIS technology potentially had to offer. The maintenance of the LAMIS hub file was transferred in 1991 to the planning department as the most active user of the LAMIS legacy. In 1992, the project management group conducted a study of the potential for GIS within the council and recommended that priority be given initially to supporting LLC and the land terrier (the record of council-owned land and related property interests). Following open tender, the council selected ESRI as the GIS supplier and implemented networked solutions for the legal department (for LLC with access also from the planning and highways departments) and for the Leeds Development Agency (who manage the council's property).

Stage 4 (1996 to 2002) — Spurred on by these early examples of the application of GIS and with the increasing availability of powerful PCs and PC-based GIS systems, many departments began to set up GIS systems. In the absence of a strong corporate strategy for GIS (and encouraged in part by the rejection of the LAMIS approach to a grand design), systems from a variety of suppliers began to be procured. None of these systems has been supported by IT services in view of the concentration of their resources on the priority Phase 1 GIS projects. However coordination has informally occurred as departments have been encouraged to pro-cure systems from the same vendor in order to potentially simplify future support and the transfer of skills and experience across departments. This has led to the predominant use of ESRI's products across the council, with some limited use of MapInfo and GGP. With the transfer of the data from the LAMIS hub file into the council's LPG, the council has begun to build upon its investment.

As a consequence of a review conducted by external consultants (2001), Leeds City Council adopted a single supplier policy for obtaining GIS applications and has committed itself to setting up a corporate gazetteer with intranet and Internet access as a necessary precursor of a corporate data management strategy. The city council is also addressing the cultural and organizational issues that influence such an approach.

11.3 WHAT ORGANIZATION HAS IT SET UP?

The development, implementation, and operation of GIS, and related land and property systems, within the council are steered by two major groups:

- **Business, Information Technology, and Strategy Group (BITS)** — now renamed the e-Council, which is the prime forum for prioritizing the allocation of funding and staff for all significant ICT developments. This is an interdepartmental group with representation at a very senior level (assistant director and above).
- **Program Board** under the e-Council, which has been established to carry forward GIS Phase 2 at a senior level.

An informal ArcView User Group was also established a number of years ago and is being reconstituted as a GIS User Group reporting on user matters to the program board.

In terms of support for GIS, an officer within IT services was specifically appointed to assist with the implementation of GIS for the LLC and land terrier functions. Apart from this, there is no formal support for GIS within the council, although some departments have built up GIS expertise within their existing staff (e.g., within planning, environment, housing, and highways departments).

11.4 WHAT DOES LEEDS CITY COUNCIL PLAN TO DO IN THE FUTURE?

For the future, the major emphasis is upon completing the implementation of city's LPG so that it is able to contribute data to the NLPG. This includes restructuring the city's gazetteer so that it conforms to BS7666 (2000) and developing adequate routines for updating and maintenance purposes.

A comprehensive GIS strategy study was undertaken by external consultants during 2001. Among its recommendations (which have been accepted) are the adoption of a single supplier for GIS applications and the creation of a central corporate gazetteer which is available to all departments via the intranet. Through utilizing the data that is held on the corporate gazetteer the aim is to enable access by the public via the Internet both to information and services.

11.5 WHAT WERE THE POSITIVE DRIVERS AND SUCCESS FACTORS FOR GIS?

Since the start of the LAMIS project in the early 1970s, there have been a number of positive drivers that have encouraged Leeds City Council in its experimentation with better spatial information:

- Early participation in the LAMIS project started the momentum toward the development of a land and property database and also provided attractive opportunities for cost and skills sharing in conjunction with its partners.
- The severe deterioration of the paper base maps that supported LLC and the land terrier provided the impetus to digitize the base maps rather than engage in the task of redrawing the paper maps every few years (only to have to put the data into a GIS that would eventually become inevitable).
- Increasing pressure to improve information handling coupled with the availability of easy-to-use GIS products tipped the balance in favor of getting hands-on experience.

In addition, a number of contributory success factors have provided a climate of encouragement for GIS within the organization:

- The influence at appropriate moments of champions for change — in particular the director of finance and other key senior staff
- The reduced storage costs since the 1970s and faster computer processing speeds that have lowered entry levels into GIS
- Increased pressure from central government for local government to get its act together and deliver seamless services, which has moved GIS and spatial data management up on the corporate agenda
- Perhaps most important of all, the concluding of the OS SLA in 1994 from which date digital maps became available to local authorities "as of right," following the payment of the negotiated annual service charge

But despite these many potential positive factors, Leeds City Council is only just beginning to be haltingly successful with GIS, and the reasons for this are discussed in the section below. The recent GIS strategy study that has been undertaken by consultants may well form the turning point for the authority to resurrect a coordinated approach to GIS and associated land and property systems in the future.

11.6 WHAT WERE THE NEGATIVE FACTORS THAT THREATENED SUCCESS?

On balance, the history of the use of spatial information within Leeds City Council is one of limited benefits that have occurred in specialized operational areas. It is only recently that the authority is beginning to capitalize upon what GIS technology and improved management of spatial information can offer. The reasons for this stem from the negative factors that have persisted in their influence and are only now beginning to be put aside so that the authority can move forward in adopting a corporate approach, in particular:

- The underfunding of the LAMIS project arising from an underestimate of costs (e.g., for data capture), which, together with unanticipated technical problems in complex systems, resulted in lengthened time scales.
- As a consequence, the loss of credibility of the LAMIS project which, despite a brave start in the early 1970s, was never completed. This pushed the authority into the lengthy "dark ages" of disillusionment, resulting in a widespread conviction across the authority that corporate-wide ambitious experiments in information systems could never be successful.
- In the case of desktop systems, the lack of awareness of senior managers that has affected the level of support.

11.7 WHAT HAVE BEEN THE PRACTICAL BENEFITS?

As a consequence of the problematic history of GIS and spatial data within Leeds City Council, it is difficult to assess the benefits that have actually been derived. However, three particular positive effects that have accrued from its past investment and have been mentioned by staff, are:

- The serious attention that is now being given to the quality of spatial data within the authority and the cleaning of data within related systems
- The corporate "community of interest" in the use of spatial data that has developed, which is not evident for other computer systems (e.g., word processing) and is due largely to the graphical power of being able to visualize topical data within GIS
- The emergence of map location as the common currency for exchanging data across departments

11.8 WHAT ARE THE LESSONS FOR OTHERS?

As one of the earliest attempts to introduce a corporate approach and infrastructure for land and property data, which dates back to the early 1970s, there are a number of key lessons that can be drawn out for other local authorities:

- The high risk of failure of very ambitious multi-faceted corporate projects unless they are carefully managed, adequately funded, and designed to deliver early demonstrable and highly visible benefits.
- The fundamental importance of the concept of a land and property hub file, which (despite the problems with LAMIS) has survived, and the data has been updated, transformed, and transferred into the council's current LPG. Early experiments in the development of gazetteers in the 1970s (which include those in Bradford and Coventry) are perhaps underrecognized in terms of their major contributions to the ultimate emergence of BS7666.
- The need for individual and organizational tenacity in achieving a coordinated framework for the management of spatial data, which (even without a strong corporate approach) is justified by the pressure to achieve joined-up services and which has to be underpinned by the ability to exchange data between departments, defined to common standards.

CHAPTER 12

Case Study — Newcastle City Council

NEWCASTLE CITY COUNCIL AT A GLANCE

Key Facts

Local authority name: Newcastle City Council
Local authority type: Metropolitan borough
Population: 270,000
Current state of operation of GIS: Single-supplier/Authority-wide GIS
Main GIS products in use: ESRI's ArcInfo (12 licenses); ArcView (88 licenses) now replaced with ArcGIS, ArcIMS, ArcSDE, and ArcPad
Applications: Map production, gazetteer, Envirocall (environmental call center), local land charges, development control, building control, grounds maintenance, street lighting, pollution monitoring, contaminated land, crime and disorder analysis, housing, business, and residential properties database
Land and Property Gazetteer status: Corporate gazetteer implemented in 1997 as integral part of local land charges system and based on standards previously established across Tyne and Wear authorities, with UPRNs held in a structured (non-BS7666) format but able to export data in BS7666 format. Migration to BS7666 accredited gazetteer completed in January 2001
GIM/GIS strategy status: Corporate GI strategy (approved in 1997, revised in 1999, and new one approved in December 2002)
Forum for steering GIS: GI strategy group that meets as needed to discuss strategy, standards, and funding issues with GIS user group at working level
Staffing for GIS: GIS team of seven staff (within IT Newcastle) with GI liaison officers in each department
Contact details: ICT business manager (GIS) (telephone 0191 2116447)

What Makes Newcastle City Council Distinctive?

Much of Newcastle's success with GIS derives directly from the strong corporate approach that it has adopted in relation to spatial information, which is backed up by an agreed corporate GI strategy, and associated standards. Of particular interest is the way that Newcastle has promoted the benefits of GIS by identifying and developing a "killer application" that has strong backing from members and officers. Not only has the Envirocall project has been able to show how improved customer relations management can be dramatically achieved, but has also demonstrated in a topical way what GIS technology is able to offer. A major lesson from the case study for other local authorities to note is the fundamental importance of identifying a showcase project for GIS. If this project is both highly visible and successful, then it can start the momentum toward GIS that will flow through into subsequent years.

Key Stages in the Implementation of GIS

Stage 1 (1993 to 1995) — Feasibility study, procurement of GIS software, and provision of map management service.

Stage 2 (1995 to 1996) — Development and implementation of corporate gazetteer in conjunction with local land charges system.

Stage 3 (1997 to 1998) — Establishment of corporate strategy for GI and rollout of GIS to all major departments, including use of GIS to support the environmental call center (Envirocall).

Stage 4 (1999 to 2002) — A period of consolidation of the strategy and building on the early foundations. Considerable work carried out on the gazetteer. Data conversion work performed, particularly in relation to the LLC search process, and the CAPS uniform suite introduced to move toward fully integrated land and property applications. A metadata database also introduced for GI pilot experiments undertaken in Internet-based GIS and establishment of partnerships with external organizations such as the health authority and police aimed at sharing spatial information.

Stage 5 (2003 to present) — In order for Newcastle City Council to implement its new GI strategy, it will put in place:
- Access to easily understandable, accurate, and current data and metadata, within appropriate security and privacy constraints, to aid decision and policy making
- Together with appropriate direction, leadership and control, systems integration, skills, standards, support, sustained levels of funding, and technology

Positive Drivers and Success Factors for GIS

- Implementation of LLC that encouraged corporate GIS, map management, and development of gazetteer
- Best value, modernizing government, the N-initiatives, and pressure for more informed decision making, which have all emphasized the need to relate and share information about geographic locations
- Strong corporate approach
- Envirocall, which has demonstrated the power of GIS to provide high-quality, easily vizualized information and deliver improvements in business efficiency

Problems that Threatened Success

- Underestimating the time and cost involved in data conversion (with consequent delays in implementation of local land charges system and corporate gazetteer).
- Initial lack of corporate budget to fund corporate GIS infrastructure, software, data, and resources (budget has now been established).
- GI liaison officers (in each department) have to undertake duties on top of day-to-day tasks.
- Conflicting demands on IT services for support and new development not clearly prioritized.

Practical Benefits from GIS

- Improved quality and efficiency of map production service.
- Saving of one planning technician post (through improved map production).
- High-quality information for management, policy, and operational decisions.
- More efficient fulfillment of statutory obligations.
- Turnaround on local land charges improved to 3 days (from between 6 and 38 days).
- Use of gazetteer to support bulk mailing has saved £2,000 to £3,000 per annum.
- Envirocall has improved call handling (with call responses on average dealt within 1½ minutes) and reduced abandoned calls (by 50%) within environmental services department.
- Early provision of data to NLPG.

12.1 WHY WAS NEWCASTLE CITY COUNCIL CHOSEN AS A CASE STUDY?

As one of the Tyne and Wear authorities that were involved from the 1970s in the National Gazetteer Pilot Study, Newcastle City Council has a long history of using spatial information. Like a number of other local authorities reviewed in this book (e.g., Aylesbury and Powys), Newcastle City Council is a prime example of the advantages for the implementation and operation of GIS that follow from adopting a strong corporate approach. By committing to an explicit corporate GI strategy and associated standards for spatial data, the council has established a clear road map for GI to guide priorities for the allocation of resources. This has enabled the council to expand the use of GIS to wide-ranging applications across all its departments, and it is planning to open up to public access through implementation of Internet–GIS.

Particularly impressive is the way that Newcastle City Council has been able to identify the killer application. The Envirocall project (see Box 12.1) has imaginatively used GIS to support the environmental call center within the Cityworks Directorate, which deals with about 150,000 calls each year relating to rubbish collection, street cleaning, roads and footpaths, grounds maintenance, and pest control. A lesson for other local authorities to note is the importance of finding one or more showcase projects for GIS, which, if politically important, highly topical, and successful, can start the momentum toward GIS that will continue in subsequent years.

12.2 THE BACKGROUND — WHAT HAS NEWCASTLE CITY COUNCIL DONE?

Newcastle City Council is an example of the implementation of a single-supplier/authority-wide GIS, using the terminology that we introduced in Chapter 8. It has introduced the software of ESRI across all of its departments with wide-ranging applications that include map production, gazetteer, Envirocall, LLC, development control, building control, grounds maintenance, street lighting, pollution monitoring,

Box 12.1   The Envirocall Project

Replacing Compulsory Competitive Tendering, the government's best value initiative, introduced in 1997, requires councils to secure best value for taxpayers' money across their services. One of the council's first best value reviews of the customer services department identified major problems in communication, with members of the public who telephoned with environmental problems and complaints kept waiting while internal enquiries were made. In addition residents were also frustrated by the lack of an appointment system for bulk refuse collection and pest control. This growing dissatisfaction underlay Newcastle City Council's successful bid to the Department of Environment, Transport, and the Regions (DETR) for a best value project in October 1997.

Taking advantage of seed funding from the DTLR, the newly-equipped Envirocall call center with 11 staff members was opened in September 1998. For those living in Newcastle with an environmental query or request, a single widely advertised telephone number is available to call. When a resident telephones, the call agent takes down the name, address, or location, and details of the request. Using an in-house application which was built by IT services, with ESRI's MapObjects (embeddable mapping components and subsequently replaced by ArcIMS) and which accesses the council's Land and Property Gazetteer, the call agent is quickly able to display a small overview map on-screen. This provides the geographic location, ward name, and other useful information such as refuse collection days. Where appropriate, mapping is used to identify the location needing attention, e.g., if a broken street light is reported, then street light locations can be overlaid on the displayed map allowing the call agent to identify the correct one. Other features such as bus stops, traffic lights, and street furniture are also available for interrogation and enable a request for service to be entered using a linked form (which is then routed to the appropriate section for action).

The Envirocall system has dramatically improved customer services. In the first year of operation, service request calls increased by 30%, average time to deal with a call dropped to $1^1/_2$ min, and the number of abandoned calls (persons giving up and hanging up before the call was answered) was halved.

Authors' summary from Newcastle City Council internal documents.

contaminated land, crime, and disorder analysis, housing, and business and residential properties databases. Currently, 12 licenses are in use for ArcInfo (used by the gazetteer team and LLC section) and 70 licenses for ArcView (used authority-wide). It is important to note that while the corporate GI strategy is strongly based on the use of ESRI's products, it does not rule out the use of an alternative supplier, provided that a good business case can be made and effective integration achieved. However, currently there is no significant use of another supplier's GIS products.

Development and implementation of GIS has taken place over four major stages:

Stage 1 (1993 to 1995) — In 1993 the council decided to investigate the feasibility of GIS and a consultancy study was commissioned (from TerraQuest Ltd.). This provided a justification for implementing GIS, which the council accepted together with requirements specifications to be used as the basis for an invitation-to-tender to select a GIS supplier. After tendering through the European Union procedures (at that time for tenders above about £120,000, for which advertisement in the *European Journal* was required) and subsequent evaluation, benchmarking, demonstrations, and site visits, ESRI was appointed as the council's GIS supplier at the end of 1994. Following installation of hardware (server, workstation, digitizers, and plotters), the planning department began to provide a map production and management service to all departments based on use of ESRI's ArcInfo (subsequently replaced in 1997 by use of ESRI's more user-friendly ArcView product).

Stage 2 (1995 to 1996) — At the second stage in implementing GIS, the council focused its efforts on the LPG and LLC system under contract with ESRI. The gazetteer was implemented initially as an integral part of the LLC system, though it was selected so as to be capable of operating ultimately as a corporate facility and was quickly extended to other applications. The gazetteer was based on standards previously established across Tyne and Wear authorities (as part of the National Gazetteer Pilot Study) with UPRNs that are held in a structured (non-BS7666) format, though it was able to export data in BS7666 format to enable it to interface to the NLPG and NLIS. To ensure that the greatest value could be made from the gazetteer in conjunction with GIS, the council decided to captured the extents (boundaries) of properties within the gazetteer, starting with "built" properties (approximating to postal addresses) and completing the capture of land parcels several years later. A gazetteer maintenance team was established in the planning department at the outset in order to keep the gazetteer up-to-date.

The LLC system was developed by ESRI to meet Newcastle's specification (and ultimately became ESRI's first commercially available ArcLLC product). In addition to legal services, those departments (planning, highways, environmental health, and city engineers) that were consulted on the response to enquiries about departmental constraints (raised on Form CON29 as part of the local land charges search) were also provided with screens onto the new system, enabling a gradual improvement in response times to be achieved.

Stage 3 (1997 to 1998) — Following a successful trial of GIS for map production and implementation of the LLC system and LPG, the council began a period of consolidation and rollout. User numbers for GIS were substantially expanded across all departments by employing ESRI's ArcView, integrated with the council's LPG, as the easy-to-use viewer for datasets of corporate interest. A major effort went into developing applications for use by departments involved in grounds maintenance, pollution monitoring, contaminated land, crime, and disorder analysis, housing, and the business and residential properties database. While all these applications have delivered business benefits to their respective departments, perhaps the most impressive and eye-catching application developed over this period was the Envirocall project (see Box 12.1). This application alone was responsible for generating a "head of steam" for GIS, which continued to the present day. Based on the success of Envirocall, the council is keen to expand the use of GIS to provide call centers and one-stop shops for other services.

This stage was also characterized by a consolidation of GI activity, in which the local authority consciously stood back after a period of enthusiastic work and rethought where it wanted to go with GIS for the future. This resulted in the agreement of a corporate GI strategy and the establishment of the GI management group to steer its implementation. The strategy contained a strong corporate commitment to sharing information through corporate groups, and implementing a street, land, and property gazetteer for use by systems across the authority. The strategy also incorporated a set of standards for GIS, which included requirements for departments to use ESRI's ArcInfo, ArcView, and MapObjects software, and for the development of systems using ESRI's Avenue (ArcView macro language) and MapObjects (embeddable GIS components) together with Visual Basic or MS Access to provide the forms and reporting facilities. The strategy committed funding for the OS SLA, map management, corporate gazetteer construction, corporate data server, software and data, and IT support. IT services were charged under the strategy with taking a lead role in coordinating and undertaking GIS development, producing a data inventory, and providing user training.

Stage 4 (1999 to 2002) — This was a period of consolidation of the strategy and building on the early foundations. Considerable work was carried out on the gazetteer, which has been successfully submitted to the NLPG. Data conversion work was performed, particularly in relation to the LLC search process, and the Caps Uniform Suite was introduced to move toward fully integrated land and property applications. A metadata database was introduced for GI.

Faced with responding to e-government targets and improved access to information about its services, Newcastle City Council has launched a major initiative to explore how Internet-based mapping may be delivered to the public and businesses. Furthering its current focus on improved customer services, the authority piloted the use of ESRI's Internet Map Server (ArcIMS) for the delivery of mapping for car parks, tourist information, public buildings, and waste reception sites. Newcastle City Council also actively explored the establishment of partnerships with the local health authority, Northumbria Police, and private sector organizations with the aim of sharing spatial data to common standards within agreed upon privacy and security constraints.

Stage 5 (2003–ongoing) — Because over 80% of any local authority's information has some kind of geographical description, be it a property address or a location, GIS are capable of bringing together most of the data held by Newcastle City Council.

GIS have the ability to address fundamental policy and service delivery issues relating to land, property, people, and services, such as "Where?" "What?" "What if …?" and "Find and show."

Modern GIS software, cost effective hardware, and increasing requirements for electronic service delivery allow GI to play a crucial role in enabling members and staff to carry out the business of Newcastle City Council more efficiently and effectively, while at the same time ensuring that the requirements of citizens and the objectives of e-government are fully met.

The prevailing view at Newcastle City Council is that GIS are simply a technology. However, GIS are not just a technology; they are increasingly part of the way in which commerce, government, and academia operate as business functions. This is recognized by the fact that the government has identified GIS as a key enabler in its national strategy for e-government.

GI can provide cost-effective and efficient solutions to the problems associated with implementing change. Many business drivers would be difficult to implement without the use of GI and GIS, and applications such as Envirocall have demonstrated how GIS can play a key role in the delivery of best value services. The council aims to repeat this success across other service areas.

GIS is of little value without the availability of appropriate GI, and one of the main aims of the current strategy is to establish a corporate infrastructure that is capable of storing this information and making it available to appropriate officers. It is also the intention to build on the success of the metadata database to ensure that officers know what information is held, how up-to-date it is, and who is responsible for it.

Another key aim of the new GI strategy is to open up access to basic GIS functionality across the whole of the organization via the intranet. The same technology will be exploited to make GI available to Newcastle's citizens and businesses. It also recommends a management framework to oversee the delivery and periodic review of the strategy, prioritization of work, and authorization of the expenditure of these funds.

In order for Newcastle City Council to implement its new GI strategy, it will put in place appropriate:

- Access to easily understandable, accurate, and current data and metadata, within appropriate security and privacy constraints, to aid decision and policy making
- Direction, leadership, and control
- Systems integration
- Skills
- Standards
- Support
- Sustained levels of funding
- Technology

By delivering the new strategy, GI will help Newcastle City Council to achieve its national, regional, and local objectives efficiently and cost effectively.

12.3 WHAT ORGANIZATION HAS IT SET UP?

Following the agreement of the GI strategy in 1997, a GI strategy group was set up with responsibility for overseeing the development and implementation of the corporate strategy. The group was chaired by the assistant director of community and housing and was attended by second or third tier officers from all directorates. The group met regularly in its early stages, but is no longer in existence. However, the new GI strategy recommends an appropriate management framework be put in place, comprised of:

- **GI Champion:**
  It is important that a GI champion is established at corporate management team level to obtain and maintain the support of all directorates. The GI champion should thus ensure that Newcastle City Council meets the targets of the GI strategy.
- **Corporate GI Steering Group:**
  Comprised of heads of service and/or senior managers from the key service areas of Newcastle City Council, a corporate GI steering group needs to be reestablished. This group would oversee and approve all corporate GI expenditure.
- **Geographical Information Liaison Officer (GILO) Group:**
  A GILO group needs to be reestablished to coordinate GI in every directorate, sharing information with each other and compiling GI data survey results. It is anticipated that these officers will be required to allocate approximately 10% of their time in order to effectively carry out their roles.

- **Geographical Information User Group:**
  GI practitioners need to be identified to communicate regularly via meetings, presentations, GeoCentre, and e-mail in order to develop and implement best practice through the authority.

  Day-to-day support for GIS is provided by the corporate GIS team of seven staff within IT Newcastle (consisting of team leader, two senior development officers, two analysts/programmers and two programmers) and is responsible for:
  - Corporate strategy
  - Helping to deliver national and corporate initiatives such as modernizing government, joined-up services, and electronic service delivery
  - Research and development
  - Metadata and spatial database administration
  - Consultancy and advice on GIS, LPG interfaces, and associated data capture
  - Desktop, Web, and field systems development, maintenance, support, and administration using ArcGIS (ArcInfo and ArcView), ArcIMS, ArcPad, and ArcSDE
  - Informal training
  - Supplier liaison

12.4 WHAT DOES NEWCASTLE CITY COUNCIL PLAN TO DO IN THE FUTURE?

For the future, Newcastle City Council has an ambitious program planned for GIS and intends to:

- Concentrate effort on completing data capture for LLC
- Ensure that any new land and property systems are linked to the corporate LPG
- Provide access to maps via the council's intranet thereby making mapping available across the whole authority
- Provide map-based information over the Internet
- Further develop the Envirocall system to cover additional services
- Develop a vitality and viability model for the city, using Web-based GI technology, to aid in delivering the "going for growth" initiative
- Expand its partnerships in the use of spatial information with other authorities and agencies

12.5 WHAT WERE THE POSITIVE DRIVERS AND SUCCESS FACTORS FOR GIS?

It will be obvious from the case study so far that there have been a number of important drivers and success factors that have underpinned the progress that Newcastle City Council has made with GIS. The following positive drivers have strongly encouraging the council to get started with GIS:

- The push to implement a computerized LLC system that was the main driver from the start and that provided the foundations for a corporate approach to GIS, establishment of map management facilities, and development of the gazetteer

- Government initiatives such as best value, modernizing government, NLPG, and NLIS, together with pressure for more informed decision making, that have emphasized the need to relate and share information about geographic locations

Once the authority decided to implement GIS, two success factors that were critical in ensuring that it made real progress with GIS were:

- The strong corporate approach reinforced by an explicit GI strategy, managed and facilitated by a GI strategy group and GIS user group, and fully supported through a central GIS team and network of GI liaison officers
- Demonstrating through the Envirocall project the power of GIS to provide high-quality, easily visualized information that delivered significant improvements in service quality

12.6 WHAT WERE THE NEGATIVE FACTORS THAT THREATENED SUCCESS?

Despite the success that Newcastle City Council has made with GIS, there have been a number of negative factors that have threatened progress, in particular:

- Underestimating the time and cost involved in data conversion that impacted development and implementation of the corporate LPG and LLC system (which have taken considerably longer than expected and subsequently cost much more than anticipated). In resolving this problem, the authority retained a consultant with project management experience and brought in temporary staff.
- Initial lack of a budget to fund the provision of the corporate GIS server and to purchase corporate software and data. However, a corporate budget has now been established as a consequence of agreeing on the new corporate GI strategy.
- GI liaison officers (in each department) have, in most cases, been allocated the role on top of day-to-day duties, which has curtailed the amount of support that they are able to give. However, this issue has been addressed in the new corporate GI strategy.
- The conflicting demands upon IT *services* for support and development of new systems has meant that resources have been stretched and priorities have often been difficult to set. The aim is to resolve this by spreading expertise from the central GIS team to GI liaison officers and other users in departments.

12.7 WHAT HAVE BEEN THE PRACTICAL BENEFITS?

As a consequence of its investment in improved GI, Newcastle claims that a broad range of practical benefits have been achieved — of which the most significant are:

- Improved quality and efficiency of the map production service through the ability to provide up-to-date, seamless mapping to meet widely varying requirements and removing the need to separately maintain duplicate and inconsistent paper map sets.
- Saving of one planning technician post early in the implementation of GIS (through not filling a vacancy) due to significant improvements in map production.

- High-quality, easily visualized information for management, policy, and operational decisions, and for provision to the public and external partners.
- More efficient fulfillment of statutory obligations (e.g., Environment Act and National Street Works Register).
- Turnaround on LLC searches, which varied between 6 days at best and 38 days at worst, dramatically improved to an average of 3 days (with further reductions in response time likely to be achievable in the future).
- Use of the LPG to support bulk mailing has saved £2000 to £3000 costs per annum through improved accuracy of addresses, ease of letter production, and ability to batch letters by geographic area for the post office to deliver.
- Envirocall has significantly improved call handling, with requests for service increasing by 30% in the first year of operation, call responses on average dealt within $1^{1}/_{2}$ min, and a reduction in abandoned calls (those who get tired of waiting and hang up) by 50%.

12.8 WHAT ARE THE LESSONS FOR OTHERS?

While there are many lessons to be learned from the Newcastle Case Study (e.g., the positive drivers and success factors in Section 12.5 and the negative factors in Section 12.6), there are two of overriding importance that should be carefully noted:

- Like a number of other case studies, much of Newcastle's success has been directly attributable to the strong corporate approach it has adopted for the implementation of GIS, backed up by an explicit corporate GI strategy and associated standards for spatial data. This strategy has set the framework for unambiguous allocation of resources and determination of priorities. While making progress corporately on a complex multidepartmental implementation is still difficult, it provides a much clearer focus for action than the departmental and semicorporate implementations that are often characterized by contradictory and competing projects that rapidly dissipate the local authority's available resources.
- Newcastle provides a very powerful example of the importance of choosing one or more showcase project for GIS that are of major political and senior officer interest. The Envirocall project has not only been able to show how improved customer relations management can be dramatically achieved, but has also been able to demonstrate in a highly topical way what GIS are able to offer. If the local authority's showcase project, like Envirocall, is both highly visible and successful, then this can build up a momentum for GIS that will carry through into subsequent years.

CHAPTER 13

Case Study — Aylesbury Vale District Council

AYLESBURY VALE DISTRICT COUNCIL AT A GLANCE

Key Facts

Local authority name: Aylesbury Vale District Council
Local authority type: District council
Population: 165,000
Current state of operation of GIS: Single-supplier/Authority-wide GIS
Main GIS products in use: ESRI's ArcInfo (7 licenses), ArcView (25 licenses), Map-Explorer, and CAPS UNI-form (50 licenses)
Applications: Map production, land charges, planning application processing, building control, forward plans, property management, electoral registration, contracts management, and insurance records
Land and Property Gazetteer status: ESRI BS7666-compliant LPG operational
GIM/GIS strategy status: GIM/GIS strategy (adopted in June 1997)
Forum for steering GIS: GIS working party (reporting to information strategy working party)
Staffing for GIS: GIS project manager, GIS officer, and LPG technician
Contact details: GIS project manager (telephone 01296 585305)

What Makes Aylesbury Vale District Council Distinctive?

Aylesbury Vale District Council provides an excellent example of a local authority that has implemented GIS within the context of a strong corporate approach. While many local authorities may find this level of corporate commitment difficult to sustain, the case study identifies the many advantages that can accrue from this approach, including unambiguous setting of priorities, integration of systems, and introduction of data standards, all underpinned by the necessary staff and financial resources to make it happen.

Key Stages in the Implementation of GIS

Stage 1 (1994 to 1997) — Initial experimentation with two GIS products (FastMap and Geobuild).

Stage 2 (1997 to 1998) — Consultancy study to recommend strategy. IS/IT strategy adopted. ESRI (UK) Ltd. selected as GIS supplier.

Stage 3 (1999) — GIS, LPG, planning, property and land charges systems implemented with associated data capture. "First cut" BS7666 LPG built.

Stage 4 (2000 to 2001) — Contracts management and insurance records databases and building control system implemented with associated data capture. Electoral registration database matched to LPG.

Positive Drivers and Success Factors for GIS

- Adoption of corporate IS/IT strategy and provision of associated funding for GIS
- Dedicated project manager and project team for implementation, with secondment of key staff
- Users receptive to introduction of new technology
- Constructive relationship with GIS vendor
- Pressure to achieve e-government targets

Problems that Threatened Success

- Poor quality source data needed cleaning in advance of data capture.
- Capture of graphic planning history data took considerably longer than anticipated.
- Teething difficulties with interface between planning and land charges systems.

Practical Benefits from GIS

- Map base and associated data available with fingertip access on the desktop
- Enquiries dealt with quicker by reception desks and other users (e.g., planning, land charges, property, contract management, and building control departments) without reference to paper sources
- Planning applications registered in less time (30-minute savings per application), and accuracy of constraints check improved
- Accuracy of land charges search responses improved, with future potential to improve response times
- One position eliminated in land charges section through improved efficiency
- One position eliminated in development control as land charges searches no longer required manual check
- Resources no longer needed for updating duplicate manual map sets

---

13.1 WHY WAS AYLESBURY VALE DISTRICT COUNCIL CHOSEN AS A CASE STUDY?

In the mid-1990s Aylesbury Vale District Council began to experiment with using two different GIS products on a departmental basis. While the Ordnance Survey liaison officer was using FastMap (from Survey Supplies) for map production and management, the planning department had selected Everest (from Geobuild) for digital preparation of the graphics needed to compile the local plan. As a result of concern that the authority was dissipating its efforts on competing GIS products without a clear consensus as to future direction, a GIS strategy study was commissioned from independent consultants during early 1997.

Aylesbury Vale District Council provides an excellent example of a local authority that has implemented GIS within the context of a strong corporate approach. As a result of the consultancy study, the framework for future implementation of GIS was agreed upon within the adoption (in June 1997) of a corporate IS/IT strategy that established a strict single-supplier policy for GIS backed up by a substantial corporate budget of £1.03 million for GIS and related systems over the following 5 years. The authority moved away from FastMap and Everest with the implementation of ESRI's GIS products, which focused on ArcView on the desktop.

While many local authorities may find this level of commitment difficult to sustain, the case study identifies the considerable advantages that can accrue from this approach: covering the unambiguous setting of priorities, channeling of staff effort, integration of systems, and introduction of data standards (including establishment of the corporate LPG), all underpinned by the necessary staff and financial resources to make it happen. It is important to note that Aylesbury Vale District Council did not start from a strong corporate tradition. It moved from its early departmental experimentation with GIS to introducing a corporate approach as a result of recognition that corporate service priorities would be best served by improving the ability of departments to use and exchange spatial data on a common basis.

13.2 THE BACKGROUND — WHAT HAS AYLESBURY VALE DISTRICT COUNCIL DONE?

Aylesbury Vale District Council is an example of the implementation of a single-supplier/authority-wide GIS, using the terminology that we introduced in Chapter 8. Since June 1997, the council has strictly enforced the implementation of GIS based on the products of a single supplier as a consequence of adopting its corporate strategy.

Three different GIS software products are currently in use across the authority:

- **ArcInfo 7** (7 licenses), which is used to support the processing of LLC searches
- **ArcView 3.1** (25 licenses), which is used across all departments for a wide range of tasks including map production, planning application processing, building control, forward plans, property management, contracts management, and insurance records
- **MapExplorer 2** (unlimited free product) which is used throughout the council for "cheap and cheerful" viewing of high-interest data within the GIS, especially at customer reception desks in the new customer services centers

The development and implementation of GIS in Aylesbury Vale District Council took place over four major stages:

Stage 1 (1994 to 1997) — Initial experimentation with GIS had begun with FastMap used for map management and production by the OSLO (within the engineer's department) and Geobuild for digital preparation of the local plan.

Stage 2 (1997 to 1998) — The consultancy project to formulate the strategy for GIS and identify the early priority areas for implementation was undertaken. The justification for investment in GIS was accepted and the corporate IS/IT strategy adopted (with GIS being one of the key elements). A substantial corporate budget of £1.03 million was committed for GIS, LPG, and related systems over the next 5 years. Following tendering, ESRI (UK) Ltd. was selected as the GIS supplier.

Stage 3 (1999) — GIS software for map production and management was rolled out widely across the council, parallel with an extensive data capture program for planning, property management, and LLC. New processing systems for LLC and property management, linked to GIS, were implemented from CAPS (the local government application software house within ESRI). The council's existing planning application processing system was replaced by the CAPS uniform planning system and interfaced to GIS. As part of this stage of the project, an LPG conforming to BS7666 was implemented (using ESRI's LPG Tools) and dynamically interfaced to the planning, property management, and LLC systems (with half hourly updates of change data to the planning and property systems; and overnight updates to land charges). The "first-cut" gazetteer (57,000 records) was built for the council by the data capture contractor as a text and graphic database (using council tax, National Non-Domestic Rates, ADDRESS-POINT, and the county council's National Street Gazetteer (NSG) file). This database was enhanced in-house, resulting in a gazetteer that currently comprises 73,000 records; the database is now continually updated. Capture of other data also took place during this stage, covering land charges registrations (30,000 records), planning history (53,000 records), and land terrier (25,000 records).

Stage 4 (2000 to 2001) — Implementation was extended to other areas, in particular contracts management (grass cutting, street cleaning, and litter picking), insurance records, and the introduction of a business processing system linked to GIS for building control. A second phase of data capture was contracted out to collect the data involved with the contracts management work. Over this period the LPG was submitted to Intelligent Addressing and further work undertaken to resolve the small number of anomalies in the data that were identified.

13.3 WHAT ORGANIZATION HAS IT SET UP?

To support the implementation of the procurement phase of Stage 2 and the whole of Stage 3 of the corporate GIS strategy, the council established a GIS working party, which was an officer group with representation from each directorate, and from each of the business functions that were linked to the LPG. The GIS working party was chaired by the GIS project manager (see below) and reported to the council's information strategy working group. The role of the working party was to:

- Coordinate and facilitate the rollout of GIS throughout the authority
- Resolve priorities between different projects when resources become overstretched
- Ensure that steady progress was made with implementation of the LPG in terms of interfacing to other council systems, in accordance with the project plan

To ensure successful management of implementation of the GIS, LPG, and associated business systems, the council appointed one of its existing staff as the GIS project manager (a user, rather than a technical expert, from the planning division who was transferred to the IT division).

To support ongoing operation of the systems, the council also established a corporate GIS team (jointly managed by the GIS project manager and the information team manager within the forward plans division). The team consisted of a GIS officer (who is an experienced ArcView specialist) and an LPG technician.

13.4 WHAT DOES AYLESBURY VALE DISTRICT COUNCIL PLAN TO DO IN THE FUTURE?

The council is currently upgrading all products used within the corporate GIS solution. This is intended to result in a solution that is considerably easier to support, exploits the most recent improvements in technology, and automates the connections to both the NLPG and NLIS.

The council plans to continue the expansion of GIS, including links from the LPG, to other areas that can potentially benefit, in particular environmental health, the customer service center, council tax, and benefits.

With the increasing emphasis on e-government, the council is currently putting in place plans to publish selected data from the GIS on its Website and to encourage interactive access.

13.5 WHAT WERE THE POSITIVE DRIVERS AND SUCCESS FACTORS FOR GIS?

Much of the success of Aylesbury Vale District Council is due to a number of significant drivers and success factors that have helped to provide a climate of support for the project. The positive drivers that have been of particular importance prior to and throughout the project are:

- The far-sightedness of the authority in recognizing, early on, the long-term benefits to service delivery that would accrue from adopting a corporate strategy for GIS within the framework of an overall IS/IT strategy.
- Consequent on the adoption of the corporate IS/IT strategy, the commitment of substantial resources to the project at the outset, which enabled implementation to progress without interruption.
- More recently, the imposed need to achieve the e-government targets has further strengthened the justification for the GIS implementation program that was already underway.

The critical success factors that have underpinned the project and maintained its profile within the authority have been:

- The appointment of a dedicated project manager and project team, together with the secondment of key staff onto the project for defined periods
- Users who were receptive to the introduction of new technology, rather than scared that it may mean that their jobs were at stake
- A continuing constructive relationship with the GIS vendor with the emergence of Aylesbury Vale District Council as one of ESRI's "show sites," which has had advantages for both parties

The champions for GIS within Aylesbury Vale District Council included the chief executive, director of corporate resources, and IT manager. These individuals saw the potential benefits of GIS and were able to enlist the support of councilors in committing the authority to an ambitious program of implementation. Also key to achieving this support was the involvement of a panel of four councilors (from each of the political parties) as a sounding board for the GIS strategy while it was in the draft stage and throughout the first 2 years of implementation.

13.6 WHAT WERE THE NEGATIVE FACTORS THAT THREATENED SUCCESS?

While there have been many positive encouragements to the implementation of GIS within Aylesbury Vale District Council, there are a number of negative factors that have potentially threatened success. These include:

- Poor quality source data:
  - For local land charges it was decided to clean the data in advance of capture, as a substantial six-person-month project using in-house staff; this was considered time well spent that eased the subsequent workload.
  - The property management data were not cleaned as adequately as they should have been before capture, which resulted in considerable time and effort by in-house staff to tidy up the data once received back from the data capture company; this delayed the go-live date for this part of the solution.
  - The planning applications history data could not be effectively cleaned and difficult cases had to be captured by in-house staff rather than by the data capture company. This resulted in the capture of graphic data relating to planning history (back to 1974) being much more protracted than originally anticipated.
- Teething difficulties with the interface between the CAPS uniform planning application processing system and LLC system, which was a "one-off" development for Aylesbury Vale District Council; these were resolved by a team effort between council and ESRI staff.

13.7 WHAT HAVE BEEN THE PRACTICAL BENEFITS?

A major prerequisite underlying the early commitment to corporate GIS was that substantial benefits would accrue from implementation, particularly from improved service delivery. Now that implementation is substantially complete, the authority recognizes that it is obtaining significant and growing benefits from GIS in the following areas (some of which it has been possible to quantify):

- The electronic map base and associated map data are now available throughout the council, including reception desks in the new customer services centers. As a result, many queries can be answered by "customer-facing" staff that previously could not be done without extensive support from the "back office" and without the need to refer to poor quality paper maps away from the desk.
- Since all staff now have access to the same master copy of OS Landline data, resources are no longer devoted to maintaining duplicate map sets that were inconsistent with each other.
- Planning applications can be registered and consultation needs identified without staff having to leave their desks. This has resulted in quicker and more accurate processing (a resulting time saving of 30 minutes per application has been estimated).
- Property enquiries can be answered quickly and accurately without staff having to leave their desks — which is a considerable asset for users who are remote from paper records.
- The accuracy of responses to LLC searches has considerably improved, and there is potential to reduce response times, though this has not yet been realized.
- One staff position within the LLC section has been eliminated as a result of computerizing the service — and one staff position in the development control service has been eliminated as LLC search responses no longer have to be checked manually.
- The issue of plans to support contracts that are let by the council can now be done electronically, with many hours of staff time saved through avoiding the need to draw plans manually.
- Data from other business applications, which have been matched against the LPG, can be mapped using automated procedures. An example of this is the mapping of property attribute data from the council's housing system.

13.8 WHAT ARE THE LESSONS FOR OTHERS?

The Aylesbury Vale District Council case study demonstrates what can be achieved within the framework of a corporate approach and stresses the importance of:

- Enlisting from the outset the support of potential champions (at the chief executive and chief officer levels) and the elected members
- Making the case for adequate resources based on potential improvements in service delivery and adding strength to the case through the ability additionally to meet e-government targets
- Monitoring and publicizing the actual benefits (including savings) that have been achieved in order to sustain momentum
- Putting in place dedicated staff to take the lead on implementation
- Not underestimating the problems and complexity of the associated data capture

While many local authorities will find it difficult to achieve this level of corporate commitment, it is important to note that Aylesbury Vale District Council did not start from a strong tradition of corporate working. Much of the impetus toward a corporate approach came from the firm belief of the early champions and the subsequent recognition of councilors that the results would be real improvements in service delivery.

CHAPTER 14

Case Study — Shepway District Council

SHEPWAY DISTRICT COUNCIL AT A GLANCE

Key Facts

Local authority name: Shepway District Council
Local authority type: District council
Population: 100,000
Current state of operation of GIS: Multi-supplier/Authority-wide GIS
Main GIS products in use: ESRI's ArcInfo and ArcView; Autodesk MapGuide
Applications: Map production, gazetteer, planning control, building control, LLC, shore-line management, deprivation mapping, town center audits, rural transport, and the NLUD
Land and Property Gazetteer status: The LLPG is advanced and already providing benefits to customers
GIM/GIS strategy status: No strategy but underlying corporate acceptance of the value of GIS
Forum for steering GIS: Office technology liaison group has an overall view of IT, including GIS
Staffing for GIS: No full-time staff dedicated 100% to GIS, but several who are extensive users
Website: http://www.shepwaydc.gov.uk
Contact details: ICT manager (telephone 01303 852267)

What Makes Shepway District Council Distinctive?

Shepway is a good example of a district council implementing Web-based GIS. Building on the experience gained in establishing a digital map base, Shepway's success is the result of a corporate approach to a cost-effective vision by working in partnership with others.

Key Stages in the Implementation of GIS

Stage 1 (1994 to 1997) — Management and printing of the corporate OS map base by environmental services

Stage 2 (1997 to 1999) — Handling coastal datasets for shoreline management by construction and health services; exploration of corporate desktop GIS

Stage 3 (1998 to 1999) — Deprivation mapping and town center audit by economic development service; enquiring into the availability of Web-based GIS products

Stage 4 (1999 to present) — Deployment of corporate GIS using the council's intranet. Extensive use of GIS for LLPG maintenance, LLC, and planning

Positive Drivers and Success Factors for GIS

- OS/Local Authority SLA that prompted the setting up a digital map base
- Government initiatives, e.g., modernizing government and NLPG
- The strong corporate approach reinforced by a simple but explicit strategy and joint working
- Reduced costs facilitated by Web-based deployment

Problems that Threatened Success

- Excessive deployment costs for Stages 1 to 3 — now resolved by using Autodesk MapGuide
- High data capture costs at all stages — which is still an issue

Practical Benefits from GIS

- Improved quality and efficiency of the map management and production service
- High-performance searching for addresses and planning and building regulations applications
- The ability to link in other address-based applications throughout the organization

---

14.1 WHY WAS SHEPWAY DISTRICT COUNCIL CHOSEN AS A CASE STUDY?

Although Shepway District Council has used GIS since 1994, its use was limited in the early years to a maximum of five intensive users employing highly functional software. During 1997 and 1998, the authority recognized the immediate need to provide online access to GI for the vast majority of its staff and ultimately to make it widely available to the general public. At the end of 1998, an external consultant drew the council's attention to the availability of new, Web-based products. At the same time, the government was making it clear that it wanted councils to be able to transact their businesses electronically within a decade. Therefore, corporately adopting a view of "if it's useful, let's do it," Shepway District Council decided to be one of the first councils to make its geographic databases available online. In April 1999, development began on Shepway's intranet, and its approach to implementing Web-based GIS led to our decision to make this Kent district council a case study.

14.2 THE BACKGROUND — WHAT HAS SHEPWAY DISTRICT COUNCIL DONE?

Shepway District Council is an example of the implementation of a multi-supplier/authority-wide GIS, using the terminology that we introduced in Chapter 8 — it has introduced ESRI software for a very small number of "power users" and Autodesk MapGuide across all of its departments. One ArcInfo license is used within environmental health, planning, and building control services for corporate map management and data capture, one ArcView license in environment and street scene for shoreline management, and two ArcView licenses in the economic development service for the town center audit, the rural transport study, the NLUD, and for deprivation mapping. Further ArcView licenses are employed for LLPG maintenance and LLC use.

There are 150 concurrent users of Autodesk MapGuide throughout the authority. The system was developed jointly by the council and an external company, with the software supplied by an Autodesk reseller, Data View Solutions, who also provided the training. The software runs on a Viglen XX3 server with a 500 MHz Intel Pentium III processor, 17 GB hard drive, and 1 GB of RAM. The operating system is Microsoft Windows NT Server 4.0, and the MapGuide server runs in association with Microsoft Internet Information Server 4.0.

The major databases and processing systems currently linked to GIS are planning applications, building regulation applications, and the planning system property database using ADDRESS-POINT. This is currently being enhanced to use the LLPG.

Development and implementation of GIS has taken place over four major stages:

Stage 1 (1994 to 1997) — The SLA between local authorities and the OS to provide digital mapping prompted environmental services to obtain ESRI's ArcInfo for the management and printing of the corporate map base. During these early years, a few expert users produced high-quality printed maps that were made available to other staff on request, but there was no online access for the vast majority of staff, leading to:

- A continuing need to walk to map cabinets in order to retrieve maps
- The inability to maintain a common map base for all staff
- The use of outdated maps that were often in poor physical condition (Dean, 2000)

Stage 2 (1997 to 1998) — In 1997 construction and health services acquired a license from ESRI to use ArcView for handling coastal datasets derived from the Beachy Head to South Foreland Shoreline Management Plan, and this activity has continued up to the present time.

Also during 1997 and 1998, the authority recognized the need to deploy GIS more widely to its staff. The huge fall in the cost of desktop GIS systems, compared with the early workstation-based systems, made this more feasible than before. It was also realized that a corporate GIS system would be able to link multiple databases, bringing forward the possibility of using a corporate address database. However, there were still factors mitigating against the extension of GIS:

- Despite desktop systems being available for around £1000 per user at the time, the cost of implementing GIS for a large number of users would still be extremely high for a district council.
- The available desktop systems generally provided too much functionality for the average user to understand.
- Desktop GIS required a relatively high-specification PC.
- There were concerns that the network efficiency could be compromised by the transfer of large quantities of map data.
- Conventional client/server technology was not suitable to deliver the council's ultimate aim of making GIS functionality widely available to the public (Dean, 2000).

Stage 3 (1998 to 1999) — During 1998 the economic development service began using ArcView for general deprivation mapping, and in 1999 this was extended to town center audits, a rural transport study, and to developing the NLUD. All this work continues at present.

At the end of 1998, an external consultant drew the attention of the council to the availability of Web-based GIS products. Although they were then still in their infancy, the following features were recognized:

- The GIS functionality provided was suitable for in excess of 95% of staff.
- The cost per desktop was extremely low (a marginal cost of around £35 relative to £1000 previously).
- Data compression techniques ensured that network efficiency would not be compromised.
- Any PC capable of running a browser with a suitable plug-in would be able to access GIS, with no specialized software required.
- Centralized deployment would reduce overheads.
- Access could be given to the public via the World Wide Web (Dean, 2000).

Stage 4 (1999 to present) — In April 1999, development began on Shepway District Council's intranet. As part of this process, extensive evaluation criteria for the selection of Web-based GIS were compiled and various products evaluated. The main criteria included were:

- Straightforward import facilities from OS data and other major GIS products
- Nonproprietary programming language
- The ability to connect to multiple databases simultaneously
- Straightforward printing facilities
- The ability to split OS maps into multiple layers (Dean, 2000)

Based on these criteria, Autodesk MapGuide was selected as the product that best suited the council's purposes, and development of the GIS system commenced in July 1999. The work was undertaken by an external consultant and the Ashford/Shepway in-house IT unit. The main system was completed within a month but was not fully deployed until the implementation of the intranet in October 1999. The GIS immediately proved to be a popular application, permitting every member of staff with a PC to use the OS digital map base. ADDRESS-POINT could also be

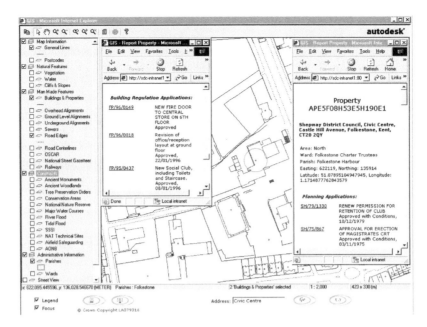

Figure 14.1 **(See Color Figure 2 following page 134.)** Shepway District Council's planning
database showing planning and building regulations applications at the Civic
Centre, Folkestone. (From Shepway District Council. Reproduced with permission
from Ordnance Survey. © Crown Copyright NC/03/16653.)

accessed via a MS SQL server database enabling around 46,000 property records
to be searched in under 3 seconds.

A batch interface to the council's planning and building control systems permits
the upload of data every evening into the central MS SQL server intranet database.
Using GIS, it is possible to find both the planning and building control applications
within any selected area. Figure 14.1 illustrates this by showing planning and
building regulations applications at the Civic Centre, Folkestone. It is intended to
extend this functionality to include the LLPG soon. Many additional coverages were
made available during 2002, including ancient monuments, ancient woodlands,
conservation areas, river and tidal flood areas, sites of special scientific interest, and
areas of outstanding natural beauty.

The council was an early participant in the NLPG project. It provided BS7666-
compliant data to the hub in December 2000 and connected the local land charges
system to the (LLPG) early in 2001. Shepway's ICT strategy indicates that major
systems should be connected to the LLPG as soon as practicable. An investigation
into the business and technical requirements involved in connecting all significant
systems to the LLPG is under way.

Shepway is already part of the NLIS at Level 2, i.e., starter system is in place,
and has found the service to be of value to some of the customers using the land
search service. An upgrade to NLIS Level, i.e., automated interface with NLIS Hub
3, is imminent.

14.3 WHAT ORGANIZATION HAS IT SET UP?

There is no formal GIS steering group, although the office technology liaison group takes an overall view of IT. In Shepway, GIS are not seen as an end in their own right but as just one, albeit important, aspect of IT that has significant potential in making corporate data more usable and accessible.

The authority is currently in the process of recruiting a systems development manager with major responsibility for corporate GIS management. This person will be responsible for ensuring that all corporate data is up-to-date, BS7666-compliant, and capable of being presented via Web-based applications to the public, staff, and councilors. The systems development manager will also assume the role of the council's OSLO.

14.4 WHAT DOES SHEPWAY DISTRICT COUNCIL PLAN TO DO IN THE FUTURE?

The aim is to provide both public and professionals with a one-stop online information service available through the Internet. As the system is rolled out to more users, the current 150-seat capacity is expected to grow to a projected capacity of around 230 seats.

The aim is to match all property-related databases with the LLPG. These will be added to GIS, allowing corporate GIS to become an umbrella for access to all spatially related data. The vision is that Shepway's residents will be able to identify a property and retrieve all the data pertaining to that property which they are entitled to view.

Shepway's Implementing Electronic Government Statement identifies the following intentions of future GIS development:

- To add additional data layers to the intranet version. Although some data have been captured and are waiting to be added to GIS, there is still a significant amount of data that could be captured. The council recognizes that capture of this data requires a coordinated effort but has not yet begun the necessary planning and resource allocation. A strategy for adding additional data layers will therefore be addressed during the financial year 2003–4. In the meantime, however, it is intended to procure expert advice on the best way forward for the capture of Shepway's properties as a polygon coverage, using the Ordnance Survey's new MasterMap product.
- Deploy GIS applications on the Internet. GIS will be used where appropriate to deliver applications having a significant spatial element. The local plan will be one of the early deployments on the Internet. As further business uses for GIS are identified, these will be added and incrementally improved in subsequent years.

The use of GIS will assist the council's operational decision making by providing officers with improved access to information and will provide an enhanced use of the Internet for customers.

14.5 WHAT WERE THE POSITIVE DRIVERS AND SUCCESS FACTORS FOR GIS?

Much of the success of Shepway District Council is due to a number of significant drivers and success factors that have helped to provide a climate of support for the project. Four positive drivers that have been of particular importance in encouraging the council to develop Web-based GIS are:

- The Ordnance Survey/Local Authority SLA in 1993 that prompted the establishment of a digital map base for the authority
- Government initiatives such as modernizing government and the NLPG that encouraged the Web-based approach
- The need to take a corporate view of spatial data and to make them available to all staff, and eventually to the public
- Reduced costs facilitated by Web-based deployment

Two success factors that have ensured that Web-based GIS is seen as an achievement by the council are:

- The strong corporate approach reinforced by a simple but explicit strategy and joint working among managers, consultants, and vendors
- Low-cost deployment opportunities for Stage 4 together with the enthusiasm of the staff

14.6 WHAT WERE THE NEGATIVE FACTORS THAT THREATENED SUCCESS?

Although there have been many positive encouragements for Shepway District Council, there are a number of negative factors that have potentially threatened success. These include:

- The excessive deployment costs for Stages 1 to 3, which were resolved by using Autodesk MapGuide corporately, thereby achieving very low per-user costs
- The high data-capture costs at all stages, which is still an issue

14.7 WHAT HAVE BEEN THE PRACTICAL BENEFITS?

GIS in Shepway have delivered a wide range of benefits:

- Improved quality and efficiency of the map management and production service — no longer necessary to use paper maps because consistently up-to-date mapping is readily available via desktop PCs
- High-performance searching for addresses, planning applications, and building regulations applications (often used in preference to the source systems)
- Ability to link in other address-based applications throughout the organization in order to give a corporate view

All these have been achieved with nonstaff costs averaging less than £10,000 annually since 1994.

14.8 WHAT ARE THE LESSONS FOR OTHERS?

The Shepway case study carries many lessons for others including:

- The advantage of a corporate approach with a clear but simple vision and a "let's do it" attitude
- The progress that can be made with an approach that is committed to action, and by working in partnership with all the other players — management, staff, vendors, consultants, and other local authorities
- The importance of striving for a cost-effective affordable (sustainable) system
- The need to take advantage of the momentum arising from government initiatives in order to develop GIS projects that deliver benefits to a wide range of users

CHAPTER 15

Case Study — London Borough of Enfield

LONDON BOROUGH OF ENFIELD AT A GLANCE

Key Facts

Local authority name: London Borough of Enfield
Local authority type: Outer London borough
Population: 265,000
Current state of operation of GIS: Multi-supplier/Authority-wide GIS in transition toward single-supplier/authority-wide GIS status
Main GIS products in use: Originally Sysdeco Records/GI3S linked to ICL PLANES, now disbanded in favor of MapInfo with text data captured in MS Access and held in Oracle 8i.
Applications: Gazetteer, development control, planning policy, LLC, environmental health, property review, and education admissions
Land and Property Gazetteer status: MapInfo BS7666 Gazetteer
GIM/GIS strategy status: Agreed corporate GIS strategy (July 2001)
Forum for steering GIS: GIS program board
Staffing for GIS: Information services manager supported by four GIS staff and three address management staff
Contact details: Group information manager (telephone 020 8379 3874)

What Makes London Borough of Enfield Distinctive?

Enfield is an example of a large London borough that has attempted, but so far has failed to achieve, a fully corporate approach to GIS in the face of strong departmentalism and lack of high-level champions. Despite having achieved only a semi-corporate approach, it is believed to have been the first local authority in the Britain to have captured all its address and property data with extents to BS7666 standards using accredited software. Like many authorities, it has had to weather the change from systems that were becoming obsolete (moving from Sysdeco Records/GI3S and ICL PLANES to systems based on MapInfo). Strong policy initiatives for joined-up government within the borough council are strengthening the willingness of departments to work together and may ultimately result in a fully corporate approach and transition into an example of single-supplier/corporate GIS.

Key Stages in the Implementation of GIS

Stage 1 (1989 to 1992) — Building of PLANES property gazetteer, with Sun network for Sysdeco records

Stage 2 (1992 to 1994) — Establishment of PLANES topic sets for planning, land charges, and property terrier

Stage 3 (1994 to 1996) — Extension of applications for contaminated land, parking control, school admissions

Stage 4 (1996 to 1999) — Maturity of semicorporate GIS, with access primarily through MapInfo on PCs

Stage 5 (2000+) — Move toward fully corporate GIS with introduction of single GIS supplier policy (MapInfo) and implementation of intranet/Internet

Positive Drivers and Success Factors for GIS

- Need to automate planning applications, land charges, and data relating to council-owned property
- Joined-up government and commitment to make council accessible to the citizen
- Commitment of staff
- Building of core team within information services unit of environmental services directorate

Problems that Threatened Success

- Strong departmentalism that impeded data sharing, with a lack of high-level champions
- Lack of understanding of core benefits of GIS, which is seen by many as a specialist technology
- Lack of flexibility of existing software
- Lack of investment
- Lack of skills

Practical Benefits from GIS

- Increased efficiency of operational processes, particularly for planning and land charges
- Improved spatial analysis for strategic management of the borough's services

15.1 WHY WAS LONDON BOROUGH OF ENFIELD CHOSEN AS A CASE STUDY?

It is believed that London Borough of Enfield was the first local authority in the country to capture all its address and property data with extents to BS7666 standards using accredited software (Sysdeco GI3S). It has a long tradition of developing and maintaining property gazetteers, starting with the implementation of ICL's PLANES database in 1989, with associated boundary data held initially in Sysdeco Records and subsequently in GI3S. With the resulting obsolescence of PLANES, and the closure of Sysdeco as a GIS supplier, the council has had to transfer its property and spatial data to systems based on MapInfo as the preferred GIS.

Despite these early initiatives in relation to land and property data, the council has attempted, but so far failed to achieve, a fully corporate approach to GIS. As in many large organizations, the introduction of an authority-wide approach has been made difficult by strong departmentalism and lack of high-level champions. Currently, the council is an example of the significant amount of progress that can be made *despite* attaining only a semicorporate approach. For the future, the situation may change as a result of strong policy initiatives for joined-up government and electronic service delivery that are strengthening the willingness of departments to work together. The result may ultimately be the adoption of a fully corporate approach within which data is shared extensively across departments through the acceptance of common standards and compatible systems.

15.2 THE BACKGROUND — WHAT HAS LONDON BOROUGH OF ENFIELD DONE?

London Borough of Enfield is an example of the implementation of a multi-supplier/authority-wide GIS, using the terminology that we introduced in Chapter 8. However, the authority is now in transition toward single-supplier/authority-wide GIS status.

In the past, it began by developing systems based on Sysdeco Records and GI3S, linked to ICL's PLANES database. MapInfo was introduced initially to coexist with Sysdeco GIS products as the user-friendly viewer of data held initially within Records and subsequently within GI3S. It has now become the GIS product that is used throughout the authority, with the major areas of application being development control, planning policy, local land charges, environmental health, property review, and education admissions.

Currently, the following MapInfo products are used by departments:

- **MapInfo Professional** (24 licenses) — which is used by environment, education, corporate, and strategic services
- **MapInfo ProViewer** — a free viewer used by 100 users
- **MapXtreme** (25 licenses) — which is MapInfo's GIS browser and which is primarily for a pilot intranet GIS application

The implementation of GIS has undergone a checkered history that involved responding to the demise of ICL PLANES as a product and Sysdeco as a supplier. London Borough of Enfield's lengthy experience of GIS can be grouped for convenience into five stages:

Stage 1 (1989 to 1992) — Initial experimentation with GIS started in 1989 with the building of the council's first property gazetteer within ICL PLANES. Sysdeco Records, running on a small network of Sun workstations within the Information Services Team (in the environmental services department), was used to capture and maintain the boundaries of properties within the gazetteer. A link was implemented between PLANES and Records to allow editing of the text and spatial data.

Stage 2 (1992 to 1994) — Following the establishment of the core gazetteer, PLANES topic sets were built up for the planning register, planning constraints, land charges, and the property terrier with associated boundaries for all but the planning register, held in Sysdeco Records.

Stage 3 (1994 to 1996) — Use of GIS was expanded through the installation of additional Sun workstations to support applications for contaminated land, parking control, and school admissions.

Stage 4 (1996 to 1999) — With the maturity of the semicorporate GIS (and directly in line with Sysdeco's product strategy), MapInfo was introduced to coexist with Sysdeco Records as a user-friendly tool for viewing and analyzing data captured and maintained within Records. Following open tender, MapInfo's BS7666 LPG was selected as the product to replace PLANES, for which ICL had, by this time, withdrawn support. Planning application boundaries began to be captured for current planning applications and enabled the rapid identification of relevant planning constraints. Over this period, the information services team took the lead in beginning the migration from Sysdeco Records to GI3S, which was heralded as Sysdeco's state-of-the-art object-oriented product.

Stage 5 (2000 to present) — Following the unexpected closure of Sysdeco as a GIS supplier, the authority committed to establishing a single GIS standard, based on MapInfo and complementary products. The "Development of a Corporate GIS and Local Street and Property Gazetteer Infrastructure" became a major council project (led by the environmental services group), which included the migration of data from PLANES and Records/GI3S and the implementation of the gazetteer to BS7666 Standards. Data matching of the council's existing property gazetteer with council tax, National Non-Domestic Rates, electoral registration, and the local street gazetteer was started to enable the accreditation of gazetteer through the IDeA. Over this period the council selected Swift, through open tender, as the supplier of new planning application processing and local land charges systems (linked to MapInfo and the gazetteer) to replace ICL's PACIS system and data held within the PLANES LLC topic set.

15.3 WHAT ORGANIZATION HAS IT SET UP?

To steer the development and implementation of GIS, the council originally set up a GIS steering group, which was in existence for a number of years. The group met only infrequently and was largely ineffective.

In practice, key decisions with regard to GIS policy have been taken by the council's IT steering group, which is composed of senior managers from each of the directorates. This group has agreed on the brief for the "Development of a Corporate GIS and Local Street and Property Gazetteer Infrastructure" project. Implementation of GIS projects has been coordinated through a GIS program board and carried out by the information services team.

Support skills for GIS are concentrated within the information services team with the environmental services group. This unit has played a major role throughout the history of GIS implementation within the borough. The unit has a split role in that it provides IT, GIS, and information skills to the whole of the environmental services directorate, but also manages the LPG and high-profile GIS datasets as a corporate service to the authority. The group information manager is supported by seven staff members with GIS skills, structured as follows:

- **GIS Team** — with team leader, three GIS assistants, and the street naming and numbering officer who are responsible for maintaining corporate GIS data and providing general support on use of systems
- **Property Information Team** — with team leader and two assistants who are responsible for street naming and numbering, LLPG maintenance, and street and property information

15.4 WHAT DOES LONDON BOROUGH OF ENFIELD PLAN TO DO IN THE FUTURE?

The challenge for London Borough of Enfield is to complete the implementation of GIS on a corporate basis so that it underpins the political priority of achieving joined-up government (see Figure 15.1). The priorities are to:

- Incorporate the GIS/GIM strategy into the council's IT strategy
- Promote the LPG as the authoritative source of address information within the council, enforcing the standard that all systems within the council are linked to it
- Develop the use of metadata and data standards

The Function of the Corporate Spatial Database

Figure 15.1 (See Color Figure 3 following page 134.) London Borough of Enfield's vision for the delivery of data from service information systems to the council's customers in conjunction with its partners. (From London Borough of Enfield.)

- Rationalize the use of street and property data
- Extend the use of GIS across departments within the framework of a clear single-supplier policy based on MapInfo
- Expand the use of Web-based GIS through the PlanWeb application, starting from a pilot project, to allow access to the LPG and core data across the intranet
- Develop pilot Internet applications
- Support "spatial enabling" projects through linkage of text databases to GIS
- Demonstrate the value of the LPG and GIS to support joined-up working so that the transition to single supplier/corporate GIS status can be fully realized

15.5 WHAT WERE THE POSITIVE DRIVERS AND SUCCESS FACTORS FOR GIS?

London Borough of Enfield has been engaged in the implementation of GIS and LPG for over 10 years. The drivers that have pushed the authority into using GIS technology have all been business oriented, the most significant being:

- Pressure to automate the processing of planning applications, which is a highly spatial activity
- Political interest in improving the processing of local land charges searches and enquiries, with the target (achieved in 1995, and subsequently) of returning search responses within 24 hours
- The need to improve the quality and availability of information relating to council-owned property, which was recognized as a major asset (with an estimated £1 billion asset value)
- The current pressure for joined-up government and electronic service delivery, emerging out of government advice, but now embodied in a political commitment to make the council more accessible to its citizens
- And most recently, the imperative of evidence-based policy to support neighbor-hood renewal and community strategies

The critical success factors that have worked with these drivers to enable the council to achieve success with GIS despite the lack of a corporate approach have been:

- The building of the core team of skills within the information services unit in the environmental services directorate
- The commitment of key staff within the core team and other departments such as corporate services, which includes the local land charges and property management functions
- The support of key directors, in particular the director of environmental services

15.6 WHAT WERE THE NEGATIVE FACTORS THAT THREATENED SUCCESS?

Despite the many positive pressures, there have been a number of significant negative factors that have held back progress. These include:

- Lack of commitment, until recently, to a corporate approach with strong departmentalism that has impeded data sharing
- Only limited appreciation of the core benefits of GIS, which is still seen by some
  as a technical problem looking for a solution
- Coupled with this, a lack of awareness of the potential impact of GIS on
  services, hence no championing by managers at a high level within the organizational structure
- Inflexibility of the early GIS and gazetteer products, resulting in inability to meet
  new needs in a responsive manner
- Erratic, rather than sustained, investment in GIS and related systems, which forced
  development to proceed in a "stop–go" manner
- Lack of skills in GIS initially, before the creation of the GIS team and GIS
  coordinator post within the information services unit, but these skills are still
  scarce and stretched

With the establishment of the council's Joined-Up Government Board, many of
these negative factors are now being overcome. The board is fostering a new vision
in which the council will be treated as a single entity, data will be seen as a corporate
resource, and services will be forced to work together. Only the long term will tell
if departmentalism truly disappears and GIS becomes recognized as a prerequisite
for integrated delivery of public services.

15.7 WHAT HAVE BEEN THE PRACTICAL BENEFITS?

In London Borough of Enfield, GIS have been used primarily to support operational processes. The emphasis has been upon improving data quality, establishing
data definition standards, and introducing formal procedures for data maintenance.
GIS have been little used to support strategic planning and monitoring. This is
because of the difficulty of getting senior officers to understand the potential role
that GIS can play and then to find the necessary resources for research and analysis.

This picture underlines the profile of benefits that have accrued to the authority:

- The greatest benefits have been directly attributable to the increase in efficiency
  of operational processes that have occurred in the processing of planning applications, local land charge searches, and property management transactions. For
  example, the most dramatic effect has been on the processing of local land charges
  searches and enquiries that are now turned around in less than 24 hours as a result
  of vastly improved access to information within GIS (for which the council
  received the Charter Mark in 1995).
- Lesser benefits have accrued from the provision of spatial analyses to support
  strategic management of the borough's services, though this is likely to be a
  growing area of return as joined-up working develops.

15.8 WHAT ARE THE LESSONS FOR OTHERS?

Looking at London Borough of Enfield's lengthy history of involvement with
GIS and LPG, there are a number of important lessons that can be drawn:

- The fundamental importance that must be placed on data, in terms of its quality, definition, and maintenance, as part of the basic infrastructure for GIS. This is an area where London Borough of Enfield has been particularly successful, and it has enabled the council to weather the storms of migrating from ICL PLANES and Sysdeco's GIS software to MapInfo as the GIS and gazetteer supplier.
- The need to conceive of GIS as a long-term program of implementation that will only be successfully achieved over a number of years and which requires the sustained commitment of resources.
- The vital precondition for success of having a core team of staff with the necessary skills and project leadership ability to make things happen.
- The need to take advantage of the momentum arising from government pressure, advice, and initiatives in order to align GIS projects so that they will deliver benefits that are widely appreciated as supporting the political and service priorities.
- The difficulty of achieving a fully corporate approach, but the many advantages of striving for it as a long-term aim.

Other authorities that view themselves as corporate would do well to review the lessons and experience of London Borough of Enfield and the pragmatic step-by-step approach that it has taken to put in place a successful GIS environment.

CHAPTER 16

Case Study — London Borough of Harrow

LONDON BOROUGH OF HARROW AT A GLANCE

Key Facts

Local authority name: London Borough of Harrow
Local authority type: London borough
Population: 220,000
Current state of operation of GIS: Multi-supplier/Authority-wide GIS
Main GIS products in use: GGP (17 licenses); SIA DataMap (10 licenses); MapInfo (3 licenses); ArcView (2 licenses); Map Shore — bundled in with SASPAC census package (1 license)
Applications: Map production, drug misuse, crime analysis, deprivation indicators, education admissions, National Street Gazetteer, property, planning applications, and land charges
Land and Property Gazetteer status: Existing Ocella gazetteer (not BS7666-compliant)
GIM/GIS strategy status: None
Forum for steering GIS: None
Staffing for GIS: No formal support for GIS within the council. Informal support from GIS development officer in chief executive's policy unit whose focus is on community safety
Contact details: GIS development officer (telephone 020 8424 1545)

What Makes London Borough of Harrow Distinctive?

London Borough of Harrow is an excellent example of an authority that has achieved considerable success in the use of GIS, despite being handicapped by the lack of a corporate approach, steering group, overall strategy, and associated budget. It has adopted a bottom-up approach to implementing GIS, which has been driven by the burning political issues of drug misuse and crime and disorder. This has encouraged data sharing in partnership with external organizations such as the police, probation service, and health authority.

Key Stages in the Implementation of GIS

Stage 1 (1989 to 1995) — Engineer's, architecture, and planning departments purchase three GGP licenses that are used for map production purposes.

Stage 2 (1996 to 1997) — Expansion of availability of GGP to other council departments (further ten licenses).

Stage 3 (1998) — Links from GGP to planning and land charges systems created. Chief executive's department starts to use GIS for community safety, emergency planning, and insurance and risk management.

Stage 4 (1999 to 2002) — GIS development officer post created in chief executive's policy unit to support work on drugs misuse and crime and disorder. Extensive information sharing and analysis in partnership with other organizations provides recognition of the value of GIS and generates ongoing commitment.

Positive Drivers and Success Factors for GIS

- Support of the Drugs Action Team (DAT) and creation of a local substance misuse database
- Legislation relating to Crime and Disorder Act and National Street Gazetteer
- Visual impact of maps displaying overlapping datasets (e.g., crime and deprivation)
- Good will of users
- Successful home office bid (CCTV and burglary reduction)
- Enthusiasm developed through wide distribution of *GIS Harrow* newsletter

Problems that Threatened Success

- Lack of GIS/GIM strategy
- Lack of budget
- No real corporate client
- No organization for steering and managing GIS

Practical Benefits from GIS

- More efficient and effective ways of working
- Closer collaboration of departments and external organizations on crime and disorder issues
- Recognition within the council, in the local press, and in professional publications

---

16.1 WHY WAS LONDON BOROUGH OF HARROW CHOSEN AS A CASE STUDY?

While some authorities have been able to implement GIS corporately from the top down, putting into place their strategic vision in a planned step-by-step manner, many authorities have not had this comparative luxury. London Borough of Harrow is an example of an authority that has implemented GIS by stealth. It has used the burning political issues of drug misuse and crime and disorder to demonstrate the role that GI can play in enabling the local authority and external partners such as the police, probation service, and health authority to assess the severity and distribution of problems as the basis for developing programs of action. Those that have been able to adopt a corporate approach have tended to focus on the technology and associated spatial data, with implementation projects prioritized so as to demonstrate the business benefits and gain commitment. Harrow's approach is the reverse: It starts with the political issues of importance to the authority and innocuously puts together the technology and spatial information from which greater insight into the problems, and debate with its partners on collaborative actions, have emerged. Harrow has also demonstrated by example how data sharing can be practically encouraged with external organizations with which the council needs to collaborate.

Local authorities without a strong corporate tradition should look closely at how Harrow has achieved recognition of the value of GIS. They should review the hot issues that currently consume the attention of their councilors, senior managers, external pressure groups, and the media, and consider whether these can be used as the basis on which GIS can "cut its teeth."

16.2 THE BACKGROUND — WHAT HAS LONDON BOROUGH OF HARROW DONE?

London Borough of Harrow is an example of the implementation of a multi-supplier/authority-wide GIS, using the terminology that we introduced in Chapter 8. While many of the departments of the council are using GIS, it is without the framework of an explicit corporate approach. Five different GIS software products are in use across the authority:

- **GGP** (17 licenses — used by about 80 people), which is the most predominantly used GIS software and which equates, by default, to being the corporate product, having been used since 1989. However, there is no single-supplier corporate policy restricting departments to use GGP, and departments are able to choose whatever GIS they consider most suitable for their specific purpose. GGP is used widely for map production, storage of user overlays of information, and also provides a "front end" to the Ocella planning and land charges systems.
- **SIA DataMap** (10 licenses), which is used for education admissions.
- **MapInfo** (3 licenses), which has been used by the chief executive's policy unit since 1999 for crime and census analysis (in conjunction with GGP within which much historical data and data of other departments is stored).
- **ArcView** (2 licenses) for National Street Gazetteer.
- **Map Shore** (1 license) for map display of census data (bundled within the SASPAC 1991 Census package).

Development and implementation of GIS has taken place over four major stages:

Stage 1 (1989 to 1995) — The engineer's, architecture, and planning departments purchased GGP (3 licenses) that were used for map production purposes.
Stage 2 (1996 to 1997) — The use of GGP was expanded to all council departments (further 10 licenses) for map production purposes.
Stage 3 (1998) — Links from GGP to Ocella's planning and land charges systems were implemented, and the use of GGP for overlay creation and spatial analysis

began to expand. As GIS skills and staff resources were limited, the chief executive's policy unit employed a student placement from University of Hertfordshire to develop the use of GGP for community safety, emergency planning, and insurance and risk management. As a result of collaboration with the police, the policy unit received detailed crime data, which was cleaned and depersonalized for use within the authority, using the Omnidata package.

Stage 4 (1999 to 2002) — In order to respond to the requirements of the Crime and Disorder Act, 1998, the post of GIS development officer was created in the chief executive's policy unit. Three MapInfo licenses were purchased to support the mapping and analysis of interagency data (particularly from the police, probation service, and health authority). Extensive analyses were undertaken (both in GGP and MapInfo) to support the work of the DAT (Drugs Action Team — see Section 16.3), and provided recognition of the value of GIS, generating ongoing commitment for collaboration.

Use of the Ocella property database has enabled standardization of the definition of "land and property units" within the planning and land charges systems. The database is not BS7666-compliant, and the council does not currently have any firm plans for upgrading or replacing it.

16.3 WHAT ORGANIZATION HAS IT SET UP?

The London Borough of Harrow does not have a forum for steering the development, implementation, and operation of GIS at a corporate level.

In the absence of a strong corporate approach, priorities and standards for GIS are decided by users on an *ad hoc* basis or within the context of working groups that are set up for specific projects.

One very important example of a working group that has acted as a catalyst for the recognition of the potential of GIS is the Harrow DAT (Drugs Action Team) Information Group that was established in February 1998 as a strategic partnership. The group was charged with developing and maintaining a local substance misuse database, as required by the white paper *Tackling Drugs Together*, with the objective of assessing the nature and scale of local drug problems in Harrow. This remit was subsequently extended (see Box 16.1) to encompass the information requirements of the new Crime and Disorder Act, 1998. The information group has documented

Box 16.1   Objectives of the Harrow DAT Information Group

1. To geographically describe the main crime and substance misuse problems in Harrow
2. To enable all partners to use the geographical representation of data to supplement their existing knowledge and expertise
3. To provide a resource where everyone involved can look at the data in the same format
4. To enable problems and solutions to be defined more easily
5. To aid in the allocation of resources
6. To set up regular, systematic analysis of chosen datasets and distribute this information to relevant groups and agencies as appropriate

Source: From Corporate Policy Support Unit, London Borough of Harrow, 2000.

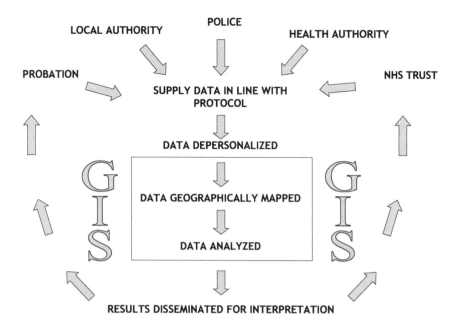

Figure 16.1   Making technology work. (*Source:* From London Borough of Harrow.)

a formal data sharing protocol that sets out the principles, procedures, and security arrangements that must be followed when data is shared between the partnership organizations. In this group, the partners collaborate on an ongoing basis to bring together and analyze data from the local authorities. Police, probation, and the health service (see Figure 16.1) have become the prime showcases for GIS within Harrow.

There is no formal support for GIS within the council. Limited informal support is provided by the GIS development officer and assistant (within the chief executive's policy unit) though their focus is upon community safety.

16.4 WHAT DOES LONDON BOROUGH OF HARROW PLAN TO DO IN THE FUTURE?

In the future, investment in GIS is likely to move away from GGP toward use of MapInfo as the preferred product, in view of the wide range of applications available for this product and its popularity among the strategic partners. In the future the DAT Information Group will be concentrating upon:

- Developing a virtual library of data available from the strategic partners, starting with a metadata catalogue.
- Improving the analysis capabilities of the strategic partners. While the datasets are mostly "skeletal" with only prototype mapping capabilities available initially, the aim is to enable skilled staff to respond more rapidly to *ad hoc* requests for analysis (e.g., analysis of police and CCTV data for crime reduction purposes).

- Expanding the ability to undertake cross-border analysis for possibly the first time in London. Harrow is in a prime position to be able to understand what affects the travel plans of criminals and the characteristics of victims and to contribute to law enforcement and crime prevention operations across administrative boundaries.
- Developing a "crime calendar" that allows investigation into the timing of criminal activity and antisocial behavior, including the effects of weather, public events, and police operations.

16.5 WHAT WERE THE POSITIVE DRIVERS AND SUCCESS FACTORS FOR GIS?

GIS has been in use in a low-key way in the London Borough of Harrow since the introduction of GGP in 1989. It has provided the means to produce maps and record simple overlay information but with little wider recognition. It is only recently that GIS has really taken off and the positive drivers for this have all arisen through legislation imposed on the authority:

- Establishment of the Drugs Action Team (DAT) and supporting DAT Information Group, following publication of the white paper *Tackling Drugs Together*
- Crime and Disorder Act, 1998, and consequent widening of the role of DAT
- Incorporation of data relating to Harrow's highway network within the National Street Gazetteer from 1995

The critical success factors that have ensured that the GIS have now been recognized as significant aids to decision making are:

- The visual impact of maps to support the work on crime and disorder relating to individual themes (e.g., census data; DETR Index of Deprivation data) and to overlapping datasets (e.g., crime data overlaid with school exclusions, deprivation data, London Ambulance Service call outs, and drugs and alcohol treatment)
- A successful bid for home office funds (for CCTV and burglary initiative) based on GIS
- The good will of users and IT at all stages
- The *GIS Harrow* newsletter (begun in February 2000), which has developed enthusiasm
- The practical demonstration, by example, of what GIS can do

16.6 WHAT WERE THE NEGATIVE FACTORS THAT THREATENED SUCCESS?

While the authority has used the strength of interest in crime and disorder to demonstrate the value of GIS, its achievements are remarkable in the absence of an overall corporate approach. These negative factors have not made success easy and include:

- The absence of a corporate steering group for GIS/GIM
- The lack of a GIS/GIM strategy
- The lack of an associated corporate budget
- No corporate client or interdepartmental sponsor for GIS

This has meant that implementation of GIS has proceeded departmentally or on a project basis (of which the Crime and Disorder Act is the outstanding example of success). But some of the benefits of working within the framework of a corporate approach will have been forgone — e.g., lack of data standards across the authority, making the interchange of data difficult at a land and property unit level.

16.7 WHAT HAVE BEEN THE PRACTICAL BENEFITS?

The most important practical benefits from GIS have been:

- More efficient and effective ways of working, e.g., processing of planning applications and land charges searches
- Closer collaboration of departments and external organizations on crime and disorder issues
- Recognition within the council, local press, and professional publications for the imaginative use of spatial information in a way that is highly relevant for analyzing and tackling social issues

16.8 WHAT ARE THE LESSONS FOR OTHERS?

Many local authorities lack a strong corporate approach, and frequently this has hindered the successful implementation of GIS. But the Harrow case study shows that it is possible to overcome this potential handicap and start to build commitment for GIS by:

- Identifying the burning political issues that are consuming the time of councilors and senior officers (most of which are spatial) and focusing on these in order to show what GIS can deliver
- Working in partnership with other departments and organizations to encourage data sharing, thereby obtaining fresh insight by access to new information about problems of common concern

This approach gains maximum impact by aligning GIS with the issues that matter. It is an approach that those attempting to implement GIS without a corporate framework should seek to emulate. Once GIS has gained a foothold, the emphasis should be redirected toward creating the corporate vision.

CHAPTER 17

Case Study — Powys County Council

POWYS COUNTY COUNCIL AT A GLANCE

Key Facts

Local authority name: Powys County Council

Local authority type: Welsh county council (unitary since local government review in April 1996)

Population: about 126,000

Current state of operation of GIS: Single-supplier/Authority-wide GIS

Main GIS products in use: MapInfo (over 100 licenses)

Applications: Map production, emergency planning, development/building control, land charges, development plan, road network/National Street Gazetteer (NSG), street works, street lighting, terrier, common land, rights of way, countryside access, census, "GIS in Schools," housing, and environmental health

Land and Property Gazetteer status: First authority to submit NSG data to OS; currently implementing BS7666 Gazetteer

GIM/GIS strategy status: Strong corporate approach with corporate GIS Aims and Mission Statement agreed from early 1990s, but with no formal GIS strategy

Forum for steering GIS: Corporate GIS steering group (with representation from all departments) that is accountable directly to strategy and resources committee

Staffing for GIS: Up to 2001 it was head of strategic planning, information and corporate GIS (in planning, economic development, and regeneration department) supported by team of three staff

Contact details: Head of research and information (telephone 01597 827511)

What Makes Powys County Council Distinctive?

Much of Powys County Council's success with GIS has been due to its very strong corporate approach since the early 1990s, which together with the enthusiasm of staff, has kept it at the forefront of technological developments. This has enabled the council to push out GIS to all departments, to all 16 high schools (gaining the AGI Local Authority Award in 1998), and to external partnerships (with RSPB, Countryside Council for Wales, Archaeological Trust, and the health authority). As Powys is the largest Welsh county by area (and the smallest by population), a major force for GIS has come from recognition of the "spatial dimension" as an important focus for bringing together information about disparate locations in order to ensure the integration of services.

Key Stages in the Implementation of GIS

Stage 1 (1991 to 1992) — Feasibility study and preparation of business case for GIS.

Stage 2 (1993 to 1994) — Approval and start of initial implementation within planning and highways departments.

Stage 3 (1994 to 1995) — Diffusion of GIS with one of three districts joining county GIS network. Establishment of external partnerships with outside agencies. Heavy investment in data capture.

Stage 4 (1996 to 1997) — GIS weathers the storm of local government reorganization in 1996. GIS installed in every high school.

Stage 5 (1998 to 2002) — Expansion, consolidation, and alignment with national initiatives such as the NSG and BS7666.

Positive Drivers and Success Factors for GIS

- Very corporate approach from the start
- Size of area covered by Powys County Council that highlighted the importance of spatial information
- Commitment and dedication of key staff
- OS SLA that promoted the availability of cost-effective digital mapping
- Enthusiasm of partner organizations for joint working
- Pressure to integrate disparate datasets

Problems that Threatened Success

- Inadequate funding for data capture, development, and rollout
- Lack of staff time; faced with competing priorities
- Low commitment from chief officers
- "Federalism" of departments that threatened the corporate approach
- Poor data especially from the OS (ADDRESS-POINT, Landline, and Boundary Line products)
- Inadequate understanding and support from GIS suppliers

Practical Benefits from GIS

- Estimated £100,000 per annum saving on map production, in particular from reduced costs of preparation of OS composites and removal of need to purchase Supply of Unpublished Survey Information (OS SUSIs)
- Effective use of visual mapping tools, e.g., for the members and the general public
- Used to justify bids for EU funds, regional assistance, rate support grant, and standard spending assessment increases, and to help set up and optimize rural rate relief scheme
- More cost-effective highway services (e.g., gritting) and improved road safety
- Improved productivity of GIS users
- 3-D capabilities for cross sections, cut and fill, line of sight, intervisibility, etc.
- Rapid reaction to spatial problems, e.g., "foot and mouth" disease

Box 17.1   Corporate GIS Implementation: Strategies, Staff, and Stakeholders

GIS fits in well with concepts of integrating data systems and adding value to information, putting it in front of decision makers at every level in an easy to understand, ready to use format, but like any other information system, GIS will give rise to changes. Truly corporate GIS implementations are still quite rare, but even small departmental GIS can have far reaching effects on organisational structure and culture of an authority. GIS implementation is about managing those changes, for example work practices, processes, information flows, management structure, and staff and organizational culture. These staff related, even social, issues are hidden and often overlooked; getting it right is as important as technical, IT, and financial success.

Traditional change management methods tend to focus on issues such as *structure, strategy, systems, skills, style,* and *shared vision,* but arguably *staff,* with their commitment, willingness, enthusiasm, inventiveness, support and promotion, are key to success. Taking that a stage further, GIS managers must also consider the wider community of *stakeholders.* Changes affecting staff and stakeholders, the social issues, are the most complex; they are also the most difficult to manage.

Source: From Gill, S. (1996) Corporate GIS Implementation: Strategies, Staff and Stakeholders, *Association for Geographic Information Conference Proceedings,* 5.2.1–5.2.8, London: AGI.

17.1 WHY WAS POWYS COUNTY COUNCIL CHOSEN AS A CASE STUDY?

Powys is a county of contrasts. It has the largest area of all the Welsh counties (2,000 square miles), together with the smallest population (126,000). While it does not have an explicit GIS strategy, it has had a very strong corporate approach to GIS since the early 1990s, with an agreed corporate mission statement and aims. In addition, the pressure from a small group of enthusiastic individuals has been particularly important in keeping it at the forefront of technological developments. A major impetus for GIS has come from the county's huge geographical area, which has encouraged a focus upon spatial information as the means to bring together information about disparate locations. A particularly impressive aspect of Powys County Council's experience is the way it has pushed out GIS into all departments, into every high school (winning the AGI Local Authority Award in 1998), and to external partnerships such as the RSPB, Countryside Council for Wales, Archaeological Trust, and health authority. Its approach to GIS has emphasized the importance of careful and professional management of spatial data, together with recognizing the potential contribution of staff and stakeholders (see Box 17.1).

17.2 THE BACKGROUND — WHAT HAS POWYS COUNTY COUNCIL DONE?

Powys County Council is an example of the implementation of a single-supplier/authority-wide GIS, using the terminology that we introduced in Chapter 8. It has introduced MapInfo software extensively throughout the council and into its high schools. It has also promoted the use of MapInfo to a number of its external partners.

Over 100 MapInfo licenses are currently in use throughout the council, with a further 16 licenses in its high schools. Uses within council departments are varied and wide ranging, including map production, development control, building control, land charges, development plan, road network and NSG, street works, street lighting, land terrier, common land, rights of way, countryside access, census analysis, and "GIS in Schools." In partnership with other organizations (e.g., RSPB, Countryside Council for Wales, Archaeological Trust, and health authority), it has forged the use of GIS as the means to exchange spatial information for policy purposes to common standards.

The checkered history of implementing and expanding the use of GIS within Powys County Council, and outward to schools and external partners, has taken place in five major stages:

Stage 1 (1991 to 1992) — In October 1991, the county council committed to a GIS pilot project for which the agreed mission was "to provide a GIS service to the Authority and to be seen as a professional and competent source of mapping and integrated information services." The key aims of the pilot project, which was championed by the head of strategic planning and information, were to implement and test an automated mapping facility, assess the ability of GIS to meet the needs of council departments, identify the main benefits likely to accrue, and promote the opportunities that GIS could potentially offer. It is worth highlighting that working in partnership with the community of Powys to obtain observations and feedback on the value of the project was recognized at the start as a fundamental part of the project. The first phase of the project was to undertake a feasibility study and to prepare the business case for implementing the pilot, including likely procurement and support costs. All departments were involved in the feasibility study; they helped to undertake a comprehensive audit of the council's map holdings and to identify potential uses for GIS. A critical assessment of all GIS products on the market was carried out, including vendor demonstrations and visits to existing user sites. It was deliberately decided not to undertake a cost-benefit analysis as part of the business case, as it was firmly believed that it would be difficult to quantify the most important benefits from GIS as these would result from improved decision making based on the availability of more consistent, accurate, and easily visualized information.

Stage 2 (1993 to 1994) — In January 1993, the finance committee and full council approved the business case and an initial revenue budget for the first stage of implementation of GIS, which aimed at assessing its potential value to the authority. A GIS steering group, comprised of representatives from the "first wave" departments (planning, highways, emergency planning, and IT support), was established to coordinate implementation and monitor progress. A reassessment of PC-based GIS products was undertaken and MapInfo selected for corporate use with 12 users initially trained in its use. By the middle of 1994, 27 licenses had been acquired and 46 users trained (in-house by GIS support staff based in the planning department). The focus to begin with was upon the use of GIS for map production purposes, exploiting the functionality of the technology for the production of paper maps and other high-quality documents incorporating maps.

Stage 3 (1994 to 1995) — The second half of 1994 through 1995 was a period of diffusion of GIS together with the establishment of partnerships. Use of GIS expanded into other departments, including rights of way, that were able to produce glossy "walk leaflets" for issuing to the public. There was also heavy investment in data capture that was undertaken in-house by staff engaged within the planning department. One of the three districts (Radnorshire) joined the county GIS network, and GIS partnerships were established with RSPB, Countryside Council for Wales, Archaeological Trust, and the health authority in order to share spatial information and ensure that it was presented for policy purposes in a common style. The public launch of GIS took place in July 1994 when GIS was taken to the Welsh National Agricultural Show (held each year within Powys) for the first time (and subsequently repeated for several years) in order to help farmers with the Integrated Administration and Control System (IACS) farming subsidy forms by providing map extracts and calculating field hectarages.

Stage 4 (1996 to 1997) — With local government reorganization in 1996, Powys became one of 22 new Welsh unitary councils, and as a consequence attention was diverted away from GIS. No specific budget was available for GIS for the first year of the new council (1996 to 1997), though eventually a £60,000 windfall was obtained to enable it to stay afloat. GIS expanded into most council departments with applications including development control, building control, development plan, street works, street lighting, land terrier, LLC, common land, rights of way, countryside access, and census. New services brought into the unitary authority from the former districts, such as housing and environmental health, also utilized GIS. Continuing the theme of diffusion begun in the earlier years, GIS was introduced into all 16 high schools (for which Powys won the AGI Local Authority Award in 1998 — see Box 17.2). The GIS manager (in the planning department) established GIS

Box 17.2   Powys Wins AGI Local Authority Award

Powys County Council has won this year's AGI Local Authority Award sponsored by the *Surveyor* magazine for the way it has extended its use of GIS into local schools.

Every high school in the region was supplied software and a CD-ROM containing Ordnance Survey base mapping and other data. The County council offered support to teachers, training at least one teacher in each school and providing ongoing advice.

Moving GIS into the school environment was not without its problems. The cost of implementation was always going to prove prohibitive but funding through the Powys Tec's science and technology programme, a special educational deal with suppliers MapInfo, and an extension of the Service Level Agreement with OS meant the project was feasible.

In the schools the take up of the new technology has been rapid and the scope impressive. "A" level pupils have used GIS in town centre and flood plain analysis but it has also been used in other subjects such as history, geology, and environmental studies.

The ease with which the GIS team has been able to implement the project has been facilitated by the corporate approach to GIS within the local authority. All information is held centrally on a computer networked to offices as far as 40 miles away. With standards common throughout, the council now has over 60 licenses and over a 100 trained users in six different departments. Extending the use of GIS beyond the boundaries of the local authority has raised awareness and there is now a desire for further licenses and a widening of the programme into junior schools.

Source: From Powys County Council Press Release, November 1998.

on a business footing by writing an SLA for corporate GIS that identified all the services that the GIS unit would provide and the costs and charges that would apply.

Stage 5 (1998 to 2002) — Since 1998, GIS within Powys County Council has continued to consolidate and expand. Use of the analysis facilities within GIS have been extended by MapInfo software add-ons that have included 3-D modeling and line-of-sight for use by the engineers and planners. Powys County Council has remained unable to contain its enthusiasm for GIS just within the authority and has actively promoted the establishment of the Welsh MapInfo User Group and AGI Cymru (both of which were initially chaired by Powys County Council's first GIS manager). Its evangelizing spirit has also led to Powys County Council promoting MapInfo to other Welsh councils, and Powys has trained some of these councils in its use. Powys County Council has taken a leading role in piloting a number of national initiatives within Wales, in particular the NSG (it was the first authority to submit its NSG to the OS), BS7666 (conformant LPG currently being implemented), Landmap (satellite imagery and digital elevation models), NLUD, and MasterMap (as part of its trialing by the OS). As part of implementing BS7666, Powys County Council has worked hard on national working groups to ensure that the LPG is relevant to rural areas and accommodates extents (boundaries of BLPUs) that may be differently defined by different users for different purposes.

17.3 WHAT ORGANIZATION HAS IT SET UP?

In order to steer the implementation and operation of GIS, Powys County Council set up, in 1993, a corporate GIS steering group, which now has representation from all departments at a senior level. The group is chaired by the corporate GIS manager within the planning department and circulates its minutes to the key chief officers.

Day-to-day support for GIS was the responsibility of the corporate GIS manager (whose main job was head of strategic planning, statistics, and information). The manager was based in the planning department and was also the council's OS liaison officer. The corporate GIS manager is ultimately accountable to the strategy and resources committee for the council's GIS services, which are financed out of the strategy and resources program area. The GIS support unit provides services to departments and external partners that are defined within an agreed SLA. The unit consists of three staff responsible to the corporate GIS manager — a corporate GIS coordinator, corporate GIS assistant, and Welsh place names officer (part-time for 18 months). Sadly, the authority was struggling to fill these important corporate GIS posts in 2002 after the skilled staff had moved on. This underlines how in demand GIS operatives now are to a wide range of organizations and how important it is to keep successful teams together.

17.4 WHAT DOES POWYS COUNTY COUNCIL PLAN TO DO IN THE FUTURE?

In April 2001, Powys County Council lost its first GIS champion (Steve Gill, former GIS manager) to the IDeA, where he now works on national e-government initiatives using GI/GIS such as NLIS, NLPG, and digital development plans on behalf of 410 local authorities. But his infectious enthusiasm for GIS lives on within the county council with the appointment of his deputy as the new GIS manager.

For the future, Powys County Council plans to continue the outward-looking approach to GIS of the past and will focus upon:

- Placing Powys's LLPG, interfaced to the NLPG, at the hub of all its databases and administrative systems
- Being able to deliver electronic services to specific client groups (starting with solicitors) through the developing NLIS
- Expanding the ability to use spatial data for strategic planning and policy formulation purposes, e.g., the use of the NLUD
- Providing desktop mapping for all staff
- Extending map-based information services to the public — in offices, libraries, and across the Internet

17.5 WHAT WERE THE POSITIVE DRIVERS AND SUCCESS FACTORS FOR GIS?

From the case study so far, it is clear that there are a number of drivers and success factors that have underpinned what Powys County Council has been able to achieve with GIS. The most significant positive drivers that have been particularly important in ensuring real progress with GIS have been:

- The commitment, dedication, and enthusiasm of key staff, including, in particular, the first GIS manager who personally championed the use of GIS, though he has widely acknowledged the significant contributions made by others within the planning and other departments, especially highways and IT. Of all the drivers this is probably the most significant.
- The recognition that in view of the large geographical area of the county, spatial information could play a major role in enabling the problems, needs, and characteristics of disparate locations to be readily visualized.
- Quick wins in key areas like planning, highways, and emergency planning.
- Service-led applications that clearly demonstrated the power and benefit of GIS.
- The concluding of the OS SLA in 1993, which promoted the availability of digital mapping "as of right," consequent upon payment of the annual fee.
- The pressure to bring together data held within separate databases to ensure improved management of services.

The critical success factors that have provided the framework for ensuring success with GIS within the organization and across its partners have been:

- The adoption of a very corporate approach to GIS from the start that has not been overly-formal in that while the aims and mission for GIS have been clearly laid out, an explicit corporate GIS strategy has not been defined. Corporate direction and priorities have been shaped up pragmatically in the light of practical experiences of what worked and was well supported, and what did not.

- The enthusiasm of the partner organizations for joint working, underpinned by the recognition that sharing topical spatial data to common standards would enable the policies and priorities of different organizations to be coordinated and reinforced.

17.6 WHAT WERE THE NEGATIVE FACTORS THAT THREATENED SUCCESS?

Despite the extremely positive progress that Powys County Council has made on GIS over the last decade, there have been a number of negative factors which have potentially threatened success along the way:

- As with almost all GIS implementations, lack of sustained funding for software, development, data capture, and rollout has been mentioned by Powys County Council as a serious negative factor that has affected the pace of implementation. This has eased as the financing of GIS has been brought under the wing of the larger and better-resourced planning department following the local government review in April 1996.
- Lack of staff time to adequately support a growing band of existing and potential GIS users who are keen to know more about it is an increasing problem. This problem is largely a direct reflection of the GIS unit's own success, and again, is characteristic of the difficult balancing act that is typical of almost all growing businesses, including other GIS implementations.
- While Powys County Council has exhibited a very strong corporate approach, it has not been without threats (successfully dealt with) from emerging departmentalism, low commitment of some chief officers, blocking tactics, and incompatible personalities.
- Poor data quality, especially from OS (e.g., ADDRESS-POINT, Landline, Boundary Line), has been an ongoing problem but has gradually been resolved as these products have improved.
- Inadequate understanding from the GIS supplier of what Powys County Council required and was trying to achieve has compromised support — which highlights that awareness building of the potential and opportunities for GIS is not just a one-way flow from supplier to client authority, but also needs to take place in the opposite direction.

17.7 WHAT HAVE BEEN THE PRACTICAL BENEFITS?

The benefits that have accrued to Powys County Council from their investment in GIS are wide-ranging and include:

- An estimated saving of about £100,000 per annum on map production attributed primarily to reduced costs of preparation of OS composites (aggregate maps for more than one conventional map sheet that are joined seamlessly across map edges) and the removal of the need to purchase OS SUSIs previously provided to local authorities as paper maps giving advance notifications of feature changes

- Significantly improved ability to justify bids for greater share of resources (e.g., EU funds, rate support grant, and standard spending assessment increases) and to help set up and optimize the rural rate relief scheme
- More cost-effective targeting of highways services (e.g., gritting) and improved road safety
- Generally improved productivity of GIS users through ease of access to consistent, high-quality spatial information to aid decision making

17.8 WHAT ARE THE LESSONS FOR OTHERS?

From the Powys case study a number of important lessons can be distilled:

- If you want to progress with GIS then choose a champion with sufficient clout to get it going and the nerve to weather the storms — but make sure that there is an understudy waiting in the wings in case the champion leaves, otherwise you will be highly vulnerable to failure.
- Ensure that you get broad support for GIS from departments, councilors, other stakeholders, and external partners — then demonstrate what GIS can offer by doing rather than just talking and planning.
- Ensure high visibility for GIS in order to build commitment by carefully choosing demonstrator projects with a high political profile, e.g., "GIS in Schools," farming subsidy applications, countryside access, and bids for EU grants.
- Highlight the importance within your organization of using spatial information as the critical means to bring together what is known about different geographic locations, which is fundamental to underpinning the whole thrust toward e-government.

PART 4

Looking to the Future

CHAPTER 18

Future Prospects and Challenges

KEY QUESTIONS AND ISSUES

- What are the future prospects for GIM in local government?
- Where is the computer revolution taking us?
- What is the likely impact of the convergence of computing and communications?
- What new technological developments will have most impact on local government?
- What will hold back the take-up of these new technologies?
- How are local authorities meeting the e-government targets?
- So what are the challenges for GIM in local government in the future?

18.1 WHAT ARE THE FUTURE PROSPECTS FOR GIM IN LOCAL GOVERNMENT?

On the face of it, the future prospects for the development of GIM in local government are extremely positive. First, there is the encouragement provided by those generalized trends that have already been identified in the earlier chapters of this book. These include:

- Continually improving computing technologies — computing capacity, wireless networking, mobile telephony, Web technology, global positioning systems, satellite imagery, and metadata
- Improved tools and techniques for spatial analysis, visualization, data search, and interoperability
- The pervasiveness of these technologies in our daily lives, providing citizens with direct access to the data they need, when they need it (e.g., location-based services, one-stop shops, and call centers)
- Political pressures stemming from the government's intention to exploit the power of information and communications technology to improve the accessibility, quality, and cost-effectiveness of public services through its e-government initiatives

- The emergence of community-based governance and greater citizen involvement both in data collection and decision making
- A wider range of concerns impinging on the individual, including health and safety, social equality, and the environment

Added to these are the recent developments in the U.K., which have a particular impact on the management of GI:

- The evolution of OS MasterMap, which in 2003 added two new intelligent layers, Integrated Transport Network (ITN) and 25 cm resolution imagery, to the existing Topography and Address layers
- The introduction of the pan-government agreement (PGA) with OS that, during its pilot year, trebled the number of central government organizations using GI in Britain from 50 to around 150, thereby promoting its wider use throughout both central and local government
- The continuing development of the LLPG initiative together with the jointly sponsored ACACIA project, which seeks to develop an integrated national infrastructure of addresses, street names, nonaddressable properties, land ownership, and other property information
- The growing use of image-based data, encouraged by improvements in high-resolution Earth observation satellites and the increasing availability of 10 cm resolution digital aerial photography covering the key U.K. towns and cities
- The continuing enhancement of the GI gateway, allowing users to search records that describe the content of sometimes very complex geospatial datasets with a new metadata creation tool called MetaGenie
- The wider availability of broadband for data transmission that assists both the development of mobile GIS and the use of the Internet, extranets, and intranets

Finally, a review of our nine case studies reveals a number of recurring messages that, if followed, would improve the prospects of GIM development within other local authorities. Although success in many of these authorities has depended on a strong corporate approach backed by an agreed upon corporate GI strategy and associated funding, there are other successful examples that have adopted a grass-roots or departmental approach to GIS development. Whichever approach is adopted, they all have cost-effective and explicit visions of what they want to achieve and can usually identify high-profile showcase projects or flagship applications. Of the other recurring positive drivers and success factors, the most important are:

- Ensuring that projects are carefully managed (preferably by a dedicated project manager), adequately funded, and designed to deliver early demonstrable and highly visible benefits
- The support of individual visionaries and champions who believe in the value of what they do and who can take others with them
- A structured and phased approach toward implementation that facilitates user involvement backed by a professional approach and both individual and organizational tenacity
- The fundamental importance of the concept of a land and property hub file and ready access to OS data

- The catalyst of achieving e-government targets
- The partnership approach, including a constructive relationship with vendors and being receptive to the ideas of users

The case studies demonstrate that a number of practical benefits flow from harnessing these positive drivers and success factors, including better quality map production, improved performance and communications, staff savings, and closer collaboration. So why, when there are all these positive factors, is there still a long way to go to achieve the full potential of GIS in U.K. local government? A look at the problems sections in each of the case studies gives part of the answer. These highlight the difficulties created by:

- The length of time spent on the capture of (often poor-quality) data
- The lack of corporate commitment and sustained funding
- The general lack of understanding of the core benefits of GIS, especially among middle and senior managers
- Both skilled staff and financial resources facing competing priorities

We will pick up these points again in Section 18.6, but first we asked Professor Michael Batty of the Centre for Advanced Spatial Analysis (CASA), University College, London, to help us assess the future for computer-based methods in local government. The next four sections are contributed by Michael and reflect his specialism in urban planning as well as his international experience. Although many of his comments focus upon urban planning, they are generally applicable to local government as a whole.

18.2 WHERE IS THE COMPUTER REVOLUTION TAKING US?

The profound thing about the computer revolution is not simply the ability to transcribe and communicate traditional media digitally and thus instantaneously. Nor is it the power to enable people to interact with one another through such media as though distance were no object. These elements are present to a greater or lesser extent in previous technologies of the industrial age, such as the telephone and telegraph. The truly profound force is the way the computer is beginning to blur boundaries between objects and ideas that were once considered entirely separate. Things that a generation or more ago were considered distinct, often sacredly so, are being juxtaposed in ways that not only blur but both excite and confuse. Culture and nature are being pushed together (machine and people, arts and sciences — the list is endless) as computers open up entirely new ways of representation, communication, interaction, prediction, and prescription.

This might sound rather grandiose in a book about the future of GI in local government, but this blurring of previously separate and distinct categories is nowhere clearer than in those domains where science and the professions are applied to areas of public interest. For a long time in urban planning, for example, there has been an explicit focus on involving the public at large not only in the assessment of plans but in the very process by which plans are prepared. Computers in planning initially divided professionals from one another and widened the gap between the public and the way plans were prepared. But as the digital revolution has deepened and broadened, computers have become an integrating mechanism. Much of this is due to their new focus on graphics and the user-friendly interfaces that they now display, although the ability to interact over the Net and the drift of computing from the desktop to the ether is as much responsible for these new possibilities. Here, we will first review what is happening within this domain. To anticipate the impact on planning and local government, it is our view that this will be very much in terms of who uses these new technologies rather than to what uses they will be put.

We consider that by the middle of the 21st century, if not sooner, most activities in everyday and professional life will be informed by digital media, and that this will open up the use of computer technologies to a very different, much wider constituency of users than there ever was in the 20th century. In this sense, the digital revolution will empower the public at large to bring data and information to each individual in a much more immediate and hence more accessible way than was ever possible hitherto. The implications of this for public planning and government are as profound as any there have ever been.

18.3 WHAT IS THE LIKELY IMPACT OF THE CONVERGENCE OF COMPUTING AND COMMUNICATIONS?

Part of this change involves the convergence of computers and communications. Once miniaturization began in earnest after the invention of the microprocessor on a chip in the early 1970s, computing became more local and more accessible, interfaces became graphical and more friendly, and many new uses emerged, all ultimately ending up on the desktop. But at the same time, computers have become devices with which to communicate. Desktops are now used as much to access information in diverse places and to send mail as to process data. In short, computers have become the devices that unlock information in diverse places and enable users to communicate quickly and efficiently over very long distances. This revolution in interactivity and the "death of distance" that is implied by such connectivity and immediacy is drifting to handheld devices where communication is wireless. The kinds of applications such devices are bringing are radically different in that users are now able to sense data in the field, capture pictures digitally, and communicate anywhere at any time with a list of potential interactions that appears endless.

This revolution will have profound implications not only for planning and local government but also for society at large. Handheld computing and wireless applications, for example, currently represent the killer application of computing in the early 21st century. At the time of writing, new low-cost wireless technologies are gathering pace in North America and threaten to overturn the same wireless technologies that are being put in place at tremendous cost for the next generation of mobile phones. Base stations that will transmit data from the Net up to 100 yards to low-cost devices can now be acquired. Such base stations can be peppered around the urban area, sensing countless activities with the prospect that this type of

Figure 18.1 (See Color Figure 4 following page 134.) New technologies: GIS on handheld devices delivering data and services.

technology is likely to turn cities into semi-intelligent sensing devices in their own right. Imagine the kinds of data that such infrastructure will transmit and what might be done with it. The prospects for wiring cities (or rather not wiring them because these technologies are wireless) and the opportunities for using this infrastructure for urban planning and management are mind-boggling. One might be skeptical of this vision, but much of it is currently happening.

In Figure 18.1, we show such handheld devices that deliver locational information about cities often in 3-D. These are being implemented in wireless fashion with links to the Web, which can be activated by a GPS that plugs into the device, thus enabling the user to walk around the city, to locate oneself, and also to pull information from the Net using other wireless plug-ins. It is a short step to even greater local interactivity as the pressure for using mobile phones for the same kinds of information access is currently demonstrating. The bottleneck in all this, of course, will be applications. Despite the technology and its interactivity, only the most routine applications will be easily developed.

However, in areas such as planning applications, in building 3-D models of the physical city, in sensing changes in the environment, and in communicating routine planning information to the public, there will be major advances (Batty et al., 2001). All this will depend upon new sensing devices, GI from many diverse sources which requires integration, and new basic software for making this data available over the Net. Many of these technologies are currently being developed not only for the desktop and handheld devices but also across the Web to which software and data is migrating. The change from stand-alone to networked computers is gradually blurring the distinction between computers and their communications, and the notion of software and data residing not actually on computers themselves (which do the main processing) but literally within local hubs — within the wires — is a prospect that promises to change the digital environment forever.

Most data that is now relevant to planning arrives in digital form. This data is unlocked through GIS or related technologies. The continued disdain for data and technology in planning, where its main educational emphasis is on procedures, is preventing the wholesale use and application of existing, nevermind new, technologies. This, as much as anything else, is a major limit to what is possible with new data and new technologies in planning. When the general public and professionals are better at using the very technologies that unlock the sectors of the new digital world than those empowered to change it, then these problems need to be seriously addressed.

18.4 WHAT NEW TECHNOLOGICAL DEVELOPMENTS WILL HAVE MOST IMPACT ON LOCAL GOVERNMENT?

We will not rehearse the multitude of applications that have been catalogued in this book, but we will identify those areas that will make a real impact on government and planning in the next decades and that are largely based on applications of new information technologies. Routine usage of IT will continue and even expand in conventional areas of data organization, survey and analysis, simulation modeling, forecasting, and related kinds of prediction. Some of these, such as those involving databases, are becoming decentralized, and there are likely to be important developments in the way such data is collected. For example, remotely sensed data on a routine basis will become more important in updating such information; while temporal data, concerned with the day-to-day control of urban activities, is already becoming essential to urban management. Apart from the obvious institutional motivation for such technology that is largely based on managerial efficiency considerations, much of the use of IT for analysis, simulation, and forecasting is predicated on individual expertise. Such developments require a new order of scientific ability among those concerned with using such tools, but current practice would suggest that such extensions are likely to be limited and will not make dramatic impacts on public planning, other than in specific, one-off instances.

The notion of planning in its strategic function based on extensive simulation and forecasting is unlikely to occur, notwithstanding selected areas of decision making being affected in this way. Although we are likely to see real-time data monitoring producing data that is fed immediately to simulation models with rapid predictive capabilities that can be acted upon for routine control, such examples are likely to be the exception rather than the rule. These advances will depend on a level of education and insight into how to use computers scientifically that is not likely to be reached. Instead, there will be a steady use and growth of such tools but in an individualistic, rather than institutional, context. It is in areas involving communication of problems and plans and ways of enabling various publics to participate that we are likely to see the greatest advances, and we will discuss these in the following sections on visualization, communication, and participation.

Visualization: In one sense, the greatest impact of the computer revolution in the last decade has been through graphics, particularly user interfaces, and more generally in ways of enabling computers to visualize numeric and qualitative data in unusual ways. Everything that can be coded can be visualized, but the greatest advances in planning have been in visualizing the environment first in abstract or map terms, and more recently in terms of the third dimension and also in more abstract ways of showing how problems and plans can be developed and evolved. The power of visualization is what makes computers so effective at communicating ideas. In terms of the physical environment, the notion of being able to see what places are like, and what they might be like, is leading to dramatic developments in being able to interact with digital versions of real environments and real plans. This area will grow dramatically.

Within 20 years, 3-D environments will be available routinely from sensed data, collected daily from various local and remotely sensed devices ranging from satellite to CCTV. From this data, real-time reconstructions will be manufactured, enabling users and participants to navigate and move within such digital environments. To give a sense of what is possible, consider the images shown in Figure 18.2 of St. Paul's Cathedral, reconstructed from remotely sensed LIDAR imagery and displayed using the desktop GIS ArcView. You can already interrogate such models within the related 3-D GIS, while you can navigate within them using the various CAD extensions that are being linked to GIS. Within a generation, every town or city will be able to produce such models, thereby showing in dramatic detail the impact of their plans.

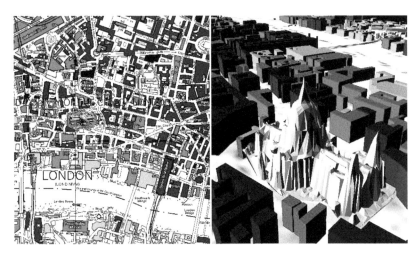

Figure 18.2 (See Color Figure 5.) The way we might visualize and navigate through digital reconstructions of real cities: An example in St. Paul's district of the city of London, using light imaging (LiDAR) data in 3-D GIS. (Reproduced with permission from Ordnance Survey. © Crown Copyright NC/03/16653.)

Communication: Visualization holds the key to effective communication, but communication is more than simply being able to disseminate information in pictorial form. The ability to interactively communicate is what the digital revolution promises. Even the current generation of mobile phones have message, picture, and related capabilities, and much routine computing in the very near future will be networked and probably wireless. We have already mentioned that digital data can be sensed automatically and communicated for eventual processing. Thus, we are beginning new ways in which data will be acquired by government. Such communication in a professional context will speed up the way different planning and control functions of local and central government will be integrated, and this might be seen as an extension of e-commerce when different authorities and groups are involved.

However, the biggest impacts are likely to be on the public at large. New ways of disseminating information are the obvious consequences of what is currently happening, and this is perhaps best seen in the plethora of community and municipal Websites that deliver planning applications information as well as the plans themselves. In a more general context, this kind of Web presence is a basis for the delivery of social and related services, again mirroring the ways in which e-commerce is developing.

Participation: In a sense, visualization and communication are the twin pillars of participation. Unlike previous technologies that seek to communicate and influence, computing is interactive; users are not simply passive receptors to be filled with information, but can act on that information, passing it back to the source. There are now many Websites devoted to such active participation, at every scale and across every kind of urban problem. In this way, Websites are being fashioned not only to disseminate information but to seek reactions whereby that information is changed and disseminated once again. In short, it is possible to see the Web, or whatever the major digital communications media in the future is called, as a means of public participation in real time. Visual technologies are key to such interaction, but so are effective interfaces that seek to entice users to act on the information. From the many such Websites available, we will show two.

In Figure 18.3A, we first show a simple environmental information system for London that enables users to extract and display information about pollution sites within London through a rudimentary query system. In Figure 18.3B, we show a page from the Hackney Building Exploratory Interactive System that enables local community users to learn about their environment in a simple but effective way and to communicate this information and their ideas back to the professionals involved. The essence of these systems is to impart information and to receive feedback, which in itself is data that informs those responsible as to the appropriateness of what they are attempting to deliver.

18.5 WHAT WILL HOLD BACK THE TAKE-UP OF THESE NEW TECHNOLOGIES?

There will be limits to these new software technologies, but the major problems that are likely to change this vision of the future depend much more on our own

A.

Figure 18.3 (See Color Figure 6.) Environmental and educational geographic information systems: **(A)** Querying pollution information at specific sites within London. (Reproduced with permission from Ordnance Survey. © Crown Copyright NC/03/16653.); **(B)** A Webpage from the Hackney Building Exploratory Interactive System for educating the public about their local environment.   *Continued.*

intrinsic values and abilities than those of technology *per se*. Despite very conscious and elaborate efforts of government, particularly central government, to develop IT as the central element in their attempts to modernize existing institutions and infrastructures, the professions, particularly those dealing with the built environment, have not embraced new trends in IT with the same fervor. There is a marked reluctance to see IT as an essential way in which ideas, information, and plans might be produced and communicated, and this lack of interest is clearly visible in education. Urban planners could be said to be the most reluctant despite the massive development of GIS in this context within the last 20 years. If the power of these technologies are to be realized in anything like the way we have indicated, then planners must be educated in their use to the point where they become advocates for these techniques. Indeed, new applications can come only from such professions. If a sustained program of education does not take place, then it is likely that the momentum will come from the private sector. For example, consider the rapid strides being made in retailing and in architectural design using computers and new kinds of data. While planners still tend to take the lead in local government, they are way behind the kind of expertise that is now available to the private sector as is witnessed in the use of simulation, modeling, and analysis, as well as in visualization that takes place in market research activities of key locational decision makers such as retailers, bankers, transport utilities, and so on.

The second issue involving take-up of these new technologies is more attitudinal than educational. The extent to which local government and the planning system might be automated depends on advocacy. This in turn comes from education at

B.

Figure 18.3 *Continued.*

least initially, but the notion of involving a wider public is something that must be intrinsic to the system itself. All that we need to say here is that the technologies of much more effective participation are now clearly available. Costs are dropping rapidly, and there are now examples of entire communities involved in using such technologies to communicate to planners. The success, or otherwise, of these ventures depends ultimately on will and interest, but there is an even more important issue: The effective use of these new tools and the new applications that might be realized depends largely upon practitioners. Therefore, in the last analysis, the kinds of visions noted here that could come to dominate the way we achieve governance and plan making in the next few decades will depend upon the attitudes and motivations of those with the professional and political responsibilities for these interests. As with all technologies, their take-up and success ultimately resides with the users, not with the machines.

18.6 HOW ARE LOCAL AUTHORITIES MEETING
E-GOVERNMENT TARGETS?

Two reports published in 2003 indicate that U.K. local authorities may fail to meet the 2005 e-government deadline. The first, by independent market analysts Datamonitor and reported in *GI News,* reveals that 36% of local authorities do not believe that they will be able to fully meet the implementation of e-government requirements by 2005. Datamonitor's survey clarifies that it is not simply a question of funds but also the need for guidance. While 92% cited funding as important, 73% said guidance is the key to success (Datamonitor, 2003).

The other survey by planning consultancy Peter Pendleton and Associates (PPA) assessed 371 planning Websites in England and Wales against 21 criteria reflecting the information and services likely to appeal most to customers. It reveals that while 89% of authorities provide easily accessible planning Web pages, just 59% have their development plans online, and only 31% have their proposals maps available to download via the Internet (PPA, 2003).

Almost two thirds of the 371 local authorities surveyed have online application registers, but only 59 enable citizens to make representations electronically, and just 32 allow users to monitor progress. Submitted application forms and the accompanying drawings and plans can be accessed on only 27 sites. Worse still, just 3% of councils offer the facility to make online planning applications. The London Borough of Wandsworth, one of the e-government pathfinder authorities, scored highest in the survey, meeting 20 out of the 21 criteria (PPA, 2003, and Johnston, 2003).

At the end of Section 18.1 we posed this question — Why, when there are all these positive factors, is there still a long way to go to achieve the full potential of GIS in U.K. local government? Returning to that question and drawing together the threads from the section on the problems experienced by the nine case studies, Michael Batty's thoughtful contribution, and the results of the above surveys, the main barriers impeding progress toward meeting GIS potential are confirmed as human and organizational rather than technical issues.

18.7 SO WHAT ARE THE CHALLENGES FOR GIM IN LOCAL
GOVERNMENT IN THE FUTURE?

The challenges are the removal of these organizational and human barriers, which we consider under the following six headings:

Raising awareness: We believe that the most important reason why the development of GIS is slow in many authorities is the continuing lack of awareness of its potential benefits. In recognition that user awareness is critical in determining the take-up of GIS, an AGI/RGS-IBG partnership developed the London Initiative to highlight the benefits, accessibility, and extraordinary usefulness of GI tools to politicians, officers, citizens, and schools. In their review of that initiative to the GIS 2000 Conference, Paul Somerfield, Chris Corbin, and Judith Mansell concluded that while the engagement of both politicians and schools had been successful, and that of the officers moderately successful, the efforts to engage the citizens had generally failed (However, their Website attracted a lot of attention and gained the involvement of many people and organizations outside the GI community.) The London Initiative also demonstrated that "preparing the nation for the Information Age places huge demands on education" (Sommerfield et al., 2000).

Turning cynicism and skepticism into enthusiasm: Adoption of new techniques does not automatically follow from awareness of them because other human and organizational difficulties can also hinder take-up. User skepticism and management reluctance to changing the old ways of working are two of the most powerful. Senior managers frequently feel threatened by the pace of change, and it is often a lack of confidence to apply technology that precludes many users. Much of the challenge for change has traditionally been cultural with legacy staff, and legacy attitudes are often at the heart of slow delivery (Chapallaz, 2001). In convincing people that change and a new way of doing business will be worthwhile, it is essential to stress the outcomes and potential rather than the jargon or technology. Changing business processes is very demanding and needs the enthusiastic commitment of users at all levels.

Managing user expectations: Citizens' expectations are being set by the developments taking place in information and communication technology. They require services tailored to their own circumstances, one-stop services that appear seamless irrespective of who provides them, choice in time, place, and medium used to access services, and not to have to repeat themselves (Brandwood, 2001). But, while increasing in numbers, not everyone can be, or indeed wants to be, an e-citizen using e-mail and the Internet. Many still want to retain the personal touch by sending letters and making visits or phone calls. Nevertheless, the potential to increase public involvement in local government is immense, especially if GIS applications are needs- and information-driven rather than technology-led. Managing user expectations, including those of local authority service providers, requires tact, dialogue, training, and partnerships.

Getting the data to everyone: Information about people, places, and movement underpins our lives and thereby the operation of local government. The "data mountain" continues to grow daily and this requires increasing storage, indexing, analysis, and searching alongside the need to cope with confidentiality, copyright, and data protection requirements. There is often a need for a culture shift from departments collecting and managing their own datasets to working with others who use similar data (Audit Scotland, 2000). Staff members are often hesitant to release their information to others, creating information silos. When this occurs, the challenge is to build bridges between these silos (Keith, 2000). The aim should be to collect data once and use it many times, delivering that information from a single point of contact. However, few have followed Wandsworth's lead in putting all but essentially confidential planning files on the Internet and using it as the principal means of interactive communication with applicants, their agents, and the public.

Meeting legislative and necessity pressures: In recent years, the business of governance has become much more challenging and difficult as expectations and the complexity of public policy issues increase within a volatile environment. Many of today's big-agenda issues are not about service delivery that is capable of being handled by single departments, but about much broader topics such as regeneration, safety, problems of the elderly or the young, green issues, sustainable development, health, crime, social exclusion, and issues involving other organizations and often focusing on particular client groups. Handling these issues, again, involves changes in both business processes and the sharing of data. Add to these the pressures of the e-government requirements, and the importance of GIS as a data gateway increases.

Sustaining the funding: The GIS "graveyard" is littered with projects that have started well but failed over time because the funding ran out or was withdrawn in budget cuts. The result is that skilled staff are lost, the system is not developed to take account of technical advances, and the data becomes out-of-date, degrading quickly. Once established, the GIS momentum must be sustainable over time. While this is not helped by the local government practice of annual budgeting, an increasing number of authorities now undertake medium-term (often 4-year) financial planning strategies. Although substantial financial and human resources are required to support and maintain a full-blown corporate GIS system, much can be achieved by refocusing existing resources that are saved as a consequence of improved GIM. Once again, we would emphasize the significance of having a long-term vision and an agreed upon strategy with ambitious but achievable aims. "Think big, start small" and "build on success" are well-worn clichés but well worth remembering.

The government intends to exploit the power of information and communication technology to improve the accessibility, quality, and cost-effectiveness of public services and to improve the relationship between citizens and those public bodies working on their behalf. That is the nub of e-government. The way ahead requires a joined-up pragmatic approach that emphasizes getting things right rather than being right, and stressing how something can be done, not why it should not be done. Despite the problems still to be overcome and the challenges to be addressed, there is huge potential for GIM in local government. The possibilities are endless; in fact, they are only limited by the imaginations of the users.

Questionnaire to Case Study
of Local Authorities

1. What is the current state of operation of GIS within your local authority? (The term *GIS* includes software that is used only for digital mapping.) Choose only one.

(a) *Corporate GIS/Single Supplier*
[only 1 GIS supplier's product(s) operational — within over 4 departments]
(b) *Corporate GIS/Multi-Supplier*
[more than 1 GIS supplier's product(s) operational — within over 4 departments]
(c) *Multi-Departmental GIS/Single Supplier*
[only 1 GIS supplier's product(s) operational — within only 2 or 3 departments]
(d) *Multi-Departmental GIS/Multi-Supplier*
[more than 1 GIS supplier's product(s) operational — within only 2 or 3 departments]
(e) *Single-Department GIS/Single Supplier*
[only 1 GIS supplier's product(s) operational — within only 1 department]
(f) *Single-Department GIS/Multi-Supplier*
[more than 1 GIS supplier's product(s) operational — within only 1 department]

Comments (optional)

2. What GIS products are currently in use within your local authority?

Product Name/ Supplier	Department/ Number of Licenses/Main Use(s)	Is the GIS Product ... (check one)		
		Fully Operational?	Partly Operational?	Being Implemented?

Comments (optional)

3. What are the major databases and processing systems that are currently linked to GIS within the local authority?

4. Has the local authority implemented a Land and Property Gazetteer? If "yes," then please provide brief details, including indicating whether it conforms to BS7666?

5. Does the local authority have an agreed GIM/GIS strategy? If "yes," then please summarize briefly the most important aspects of it (and provide a copy if available). Also please indicate whether it has been agreed by senior management and/or elected members, and how it relates to the authority's IS/IT strategies (i.e., whether it is business/policy led, or technology/process led).

6. What are the local authority's firm plans for the future implementation of GIS, and GIS-related projects (including Land and Property Gazetteer), over the next year?

7. In terms of the organization which currently exists in order to steer and support the use of GIM/GIS within the local authority ...
   (a) Is there a forum for steering the development, implementation, and operation of GIS at a high level within the local authority (e.g., GI strategy group or GIS steering committee)? If "yes," then please provide further details including a brief summary of its terms of reference, together with any supporting documents if available. If "no," then please explain how this steering role is undertaken, if at all.

   (b) Is there an officer or unit of staff (e.g., GIS officer, GIS support unit) which has responsibility for supporting departments in the use of GIS? If "yes," then please provide further details including a brief summary of responsibilities. If "no," then please explain how support is provided, if at all.

   (c) In which section/department is the Ordnance Survey liaison officer (OSLO) for the local authority (and how does the OSLO relate to any forum and support officer/unit identified in your response to 7a and 7b above)?

8. Please describe briefly in the table below the key stages in the development of use of GIS within the local authority up to the present (also provide any documentation, e.g., articles or reports).

Stage (Either just numbered consecutively or using a short "snappy" descriptive title for each stage)	Year(s) (Start–Finish)	Department(s) Involved	Activity

Comments (optional)

9. Looking at the key stages in the development of use of GIS within the above table...
   (a) What do you see as the main "drivers" which have pushed the local authority into using GIS, and within which of the above stages have they had the most impact?

(b) What are the most important positive factors which have encouraged the successful use of GIS within the local authority, and within which of the above stages have they had the most impact?

(c) What are the most serious negative factors that have threatened the successful use of GIS within the local authority, and within which of the above stages have they had the most impact?

(d) Which of these negative factors has the local authority been able to resolve, and by what actions?

(e) Which of these negative factors has the local authority not been able to resolve, and for what reasons?

10. What are the main benefits (particularly any that are tangible or have been quantified) that have occurred within the local authority through investment in GIS (please provide any supporting documents if appropriate)? (It would be helpful if you could comment on whether GIS has improved decision making, altered or improved the efficiency/effectiveness/way of working of the authority, or helped public participation and consultation — with reference to examples if appropriate.)

11. Do you have any information on the capital and revenue costs that the local authority has invested in GIS?

12. What do you think is the most notable aspect of your local authority's experiences in GIS which should be highlighted within the case study? (In this connection, it would also be helpful to know who, if anyone, were the GIS "champions" in the authority.)

13. And finally ... are there any other comments which you would like to make with regard to the use of GIS within your local authority, or more generally in relation to the theme of the book?

Many thanks for your contribution.

Useful Websites

- Association for Geographic Information (AGI) — www.agi.org.uk
- GIGateway — www.gigateway.org.uk
- HM Land Registry — www.landreg.gov.uk
- Improvement and Development Agency (IDeA) — www.idea.gov.uk
- Intra-Governmental Group on Geographic Information (IGGI) — www.iggi.gov.uk
- Local Authority Secure Electoral Register (LASER) — www.idea.gov.uk/laser
- National Land and Property Gazetteer (NLPG) — www.nlpg.org.uk
- National Land Use Database (NLUD) — www.nlud.org.uk
- National Street Gazetteer (NSG) — www.nsg.org.uk
- Office of the Deputy Prime Minister (ODPM) — www.odpm.gov.uk
- Open GIS Consortium (OGC) — www.opengis.org
- Ordnance Survey (OS) — www.ordsvygov.uk
- Planning Portal — www.planningportal.gov.uk
- Royal Institution of Chartered Surveyors (RICS) — www.rics.org.uk
- Royal Town Planning Institute (RTPI) — www.rtpi.org.uk
- UK Online Citizen Portal — www.ukonline.gov.uk
- Society of IT Managers (SOCITM) — www.socitm.gov.uk

References

Adnitt, N. (1998) Local Authority GIS — A Grassroots Approach, *AGI Conference 1998 Proceedings,* London: AGI.

Adnitt, N. (2000) e-Government, Great Ideas, Meanwhile Back in Reality … , *AGI Conference at GIS 2000 Proceedings,* t2.10, London: AGI.

AGI (1993) *GIS in Schools,* London: Association for Geographic Information.

AGI (1995) *AGI Local Government GIS Survey,* London: Association for Geographic Information.

AGI (1996) *Guidelines for Geographic Information Content and Quality,* London: Association for Geographic Information.

AGI (2002) *Activities and Achievements: Report to Members 2002,* London: Association for Geographic Information.

Allinson, J. (1994) The Breaking of the Third Wave: The Demise of GIS, *Association for Geographic Information Conference Proceedings,* 22.1.1–22.1.5, London: AGI.

Audit Commission (1988) *Local Authority Property — A Management Overview,* London: HMSO.

Audit Scotland (2000) *Common Data, Common Sense: Modernising Information Management in Councils,* Edinburgh: Audit Scotland.

Autodesk (2000) *A Strategic Guide to Enterprise GIS Decisions,* San Rafael: Autodesk.

Aybet, J. (1996) The Role of GIS in Business Process Reengineering, *Association for Geographic Information Conference Proceedings,* 2.7.1–2.7.4, London: AGI.

Baker, M. (1999) *Selecting a GIS Supplier,* presented to the RTPI GIS and IT for Planners Conference on April 23, 1999.

Ball, C.E. and Simmons, E. (1993) GIS Quality and Productivity in Development Control, in *GIS Potential and Applications,* pp. 43–65, London: Royal Town Planning Institute.

Barr, R. (2000) ASP — A Significant Progression, or a Snake in the Grass? *GI News,* vol. 1, no. 2, pp. 23–24.

Batty, M., Chapman, D., Evans, S., Haklay, M., Kueppers, S., Shiode, N., Smith, A., and Torrens, P. (2001) Visualizing the city: Communicating urban design to planners and decision-makers, in R. Brail and R. Klosterman (Ed.), *Planning Support Systems: Integrating Geographic Information Systems, Models, and Visualization Tools,* ESRI Press, Redlands, CA, 405–443.

Black, A. (2000) The National Land and Property Gazetteer (NLPG) — An Update on Work in Progress, *GI News,* vol. 1, no. 5, pp. 60–61.

Brandwood, S. (2001) How Local Government's Information Age Agenda will benefit the GI Industry, *AGI Conference at GIS 2001 Proceedings,* Workshop 3.2, London: AGI.

Bristol City Council (1999) LPG BS 7666 Information Audit: A Joined-up GIS Solution, *Report to the Management Team,* Bristol City Council.

Buchanan, H. (1997) Spatial Data: A Guide, in Green, D.R. and Rix, D. (Eds.), *AGI Source Book for Geographic Information Systems 1997,* London: AGI.

Cabinet Office (1999) *Modernising Government,* London: HMSO.

Cabinet Office (2000) *E-Government Strategy,* London: HMSO.

Cabinet Office (2000a) *Implementing e-Government: Guidelines for Local Government,* London: HMSO.

Campbell, H.J. (1996) Theoretical perspectives on the diffusion of GIS technologies, in Masser, I., Campbell, H., and Craglia, M. (Eds.) *GIS Diffusion: The Adoption and Use of Geographical Information Systems in Local Government in Europe,* London: Taylor & Francis, pp. 23–45.

Campbell, H.J. and Masser, I. (1993) The impact of geographic information systems on British local government, in Mather, P. (Ed.) *Geographic Information Handling,* Chichester: John Wiley & Sons.

Campbell, H.J. and Masser, I. (1995) *GIS and Organisations: How effective are GIS in practice?,* London: Taylor & Francis.

Chapallaz, N. (2001) Geography Matters: a Plan to Underpin e-Government, *AGI Conference at GIS 2001 Proceedings,* t2.20, London: AGI.

Chorley, R. (1991) Foreword in *Handling Geographical Information: Methodology and Potential Applications,* Masser, I. and Blakemore, M. (Eds.), Harlow, U.K.: Longman.

Chorley, R. (1997) Opening address to the Ten Years after Chorley symposium, reproduced in Appendix E of *Beyond Chorley: Current Geographic Information Issues,* London: AGI.

CMP (2000) *GIS 2000 Event Guide,* London, CMP Europe.

Coote, A. (1997) Geographical Information in the Millennium, *Proceedings of Symposium on the Future for Geographic Information: 10 Years after Chorley,* London: AGI.

Datamonitor, (2003) *Technology Opportunities in the UK Public Sector,* reported in *GI News,* vol. 3, no. 4, p. 7.

Davies, J. (1995) GIS and the Local Government Review, *Association for Geographic Information Conference Proceedings,* 6.2.1–6.2.4, London: AGI.

Dean, S. (2000) *Implementing Web-base GIS in Shepway District Council,* Proceedings of the Association for Geographic Information Conference, W3.7, London: CMP Europe.

Denniss, A. (2000) Ortho-rectified Imagery and its Uses, *GI News,* vol. 1, no. 6, pp. 39–41.

DETR (1998) *Modern Local Government: In Touch with the People,* London: HMSO.

DETR (2000) *Information Age Government: Targets for Local Government,* London: HMSO.

DOE (1972) *General Information System for Planning,* Report of the Joint Local Authority, Scottish Development Department and Department of the Environment Study Team, London: Department of the Environment.

DOE (1987) *Handling Geographic Information: Report of the Committee of Enquiry* chaired by Lord Chorley, London: HMSO.

Elliott, L. (2000) Unlocking Geospatial Information? AskGIraffe for the Key, *AGI Conference at GIS 2000 Proceedings,* t1.2, London: AGI.

England, J. (1995) Unlocking the Corporate Door, *Association for Geographic Information Conference Proceedings,* 6.4.1–6.4.5, London: AGI.

EUROGI (2000) *Towards a Strategy for Geographic Information in Europe,* Apeldoorn, Netherlands: EUROGI Secretariat.

Gault, I. and Peutherer, D. (1990) Developing GIS in Local Government in the UK: Case Studies from Birmingham and Strathclyde Regional Council, in Worrall, L. (Ed.) *Geographic Information Systems: Developments and Applications,* London: Belhaven Press, pp.109–132.

Geldermans, S. and Hoogenboom, M. (2001) The Business Case for GIS — an Absolute Necessary but Still Rare Phenomenon, *Geo Informatics,* vol. 4, no. 2, pp. 22–25.

Gilfoyle, I. and Challen, D. (1986) Digital Mapping — The Cheshire Experience, *Proceedings of Auto Carto 1986,* vol. 2, pp. 472–479, London.

Gill, S. (1996) Corporate GIS Implementation: Strategies, Staff and Stakeholders, *Association for Geographic Information Conference Proceedings,* 5.2.1–5.2.8, London: AGI.

Gill, S.G. (1998) Strategic Planning for Local Government, *Association for Geographic Information Conference Proceedings,* London: AGI.

Gill, S.G. (2000) Progress in Wales: GI at the Grass Roots, *AGI Conference at GIS 2000 Proceedings,* ps. 8, London: AGI.

Harrison, A.R. (2000) The National Land Use Database: Developing a Framework for Spatial Referencing and Classification of Land Use Feature, *AGI Conference at GIS 2000 Proceedings,* w5.6, London: AGI.

Heywood, I. (1997) *Beyond Chorley: Current Geographic Information Issues,* London: AGI.

Heywood, I., Cornelius, S., and Carver, S. (1998) *An Introduction to Geographical Information Systems,* Harlow, U.K.: Longman.

Hoggett, P. (1987) A farewell to mass production? Decentralisation as an emergent private and public sector paradigm, in Hoggett, P. and Hambleton, R. (Eds.) *Decentralisation and Democracy: Localising Public Services,* SAUS Occasional Paper 28, Bristol: SAUS, University of Bristol, pp. 215–33.

Hoogenraad, B. (2000) Intergraph: 30 Years of Innovative Technology, *GIM International,* vol. 14, no. 1, pp. 64–67.

Humphries, A. and Marlow, M. (1995) *Managing Your Address,* presented to the RTPI's IT and GIS Conference on March 24, 1995.

IDeA (2000) *IdeaOnline,* available HTTP: <http://www.idea.gov.uk/htm> (accessed January 27, 2000).

IGGI (2000) *Geographic Information: A Charter Standard Statement (GICSS),* London: Department of the Environment, Transport and the Regions.

IGGI (2000a) *The Principles of Good Data Management,* London: Department of the Environment, Transport and the Regions.

Innes, J.E. and Simpson, D.M. (1993) Implementing GIS for Planning: Lessons from the History of Technological Innovation, *American Planning Association Journal,* vol. 59, no. 2, pp. 230–236.

Johnston, B. (2003) Tools for the Electronic Age, *Planning,* no. 1536, pp. 12–13.

Keith, G. (2000) Using Intranet GIS to support Corporate GIS Service Delivery in Local Government: An Information-led Approach, *AGI Conference at GIS 2000 Proceedings,* Workshop 3.1, London: AGI.

Kendall, G. (1999) GIS Data for Business and Marketing — A Beginner's Guide, *Mapping Awareness,* vol.13, no. 11, pp. 32–34 and p. 50.

Lawrence, V.V. (1997) The Second Age of GIS, *Proceedings of Symposium on the Future for Geographic Information: 10 Years after Chorley,* London: AGI.

Lawrence, V.V. (1998) GIS: Does it Lie in the Future of Your Business? *GIS 1998 Show Preview.*

Lawrence, V.V. (2002) Government Goes Geographic, *GI News,* vol. 2, no. 6, pp. 27–29.

Lawrence, V. and Parsons, E. (1997) GIS in Disguise: Improving Decision-making with Geography, *Association for Geographic Information Conference Proceedings,* 3.2.1–3.2.5, London: AGI.

Leslie, S. (1998) Behind the Times, *Financial and Administration Journal's Special Edition for GIS' 98,* pp. 95–96.

Local Government Management Board (1991) *Spatial Information Case Study: Welwyn Hatfield District Council,* London: LGMB.

Local Government Management Board (1993) *Case Studies 93: Experiences in Geographic Information Management,* Luton: LGMB.

Local Government Management Board (1993a) *Benchmarking Guidelines — A Guide to Evaluation 'Weighing the Options',* Luton: LGMB.

Local Government Management Board (1994) *Information for Caring: GIS in Social Services,* Luton: LGMB.

Local Government Management Board (1995) *GIS: Go with the Flow,* Luton: LGMB.

Lochhead, D. (2000) GIS: Providing a Foundation for Community Governance, *AGI Conference at GIS 2000 Proceedings,* w3.4, London: AGI.

Maguire, D.J. (2001) GIS Systems without Barriers — an interview with C. Lemmens, Contributing Editor of *GIM International,* vol.15, no. 5, p. 40.

Martin, D. (1996) *Geographic Information Systems: Socio-economic applications,* 2nd ed., London: Routledge.

Masser, I., Campbell, H.J., and Craglia, M. (1996) (Eds.) *GIS Diffusion: The Adoption and Use of Geographic Information Systems in Local Government in Europe,* London: Taylor & Francis.

Masser, I. (1998) *Governments and Geographic Information,* London: Taylor & Francis.

Masser, I. and Campbell, H.J. (1994) The Take-up of GIS in Local Government: The LGMB/University of Sheffield Project, *Association for Geographic Information Conference Proceedings,* 14.2.1–14.2.6, London: AGI.

McFarlan, F.W. (1981) Portfolio Approach to Information Systems, *Harvard Business Review,* pp. 142–150.

Musgrave, T. (2000) 1000 Years of Ownership: Registering the City of Bristol, *AGI Conference at GIS 2000 Proceedings,* w3.3, London: AGI.

Nicholson, M. (2001) What the NLPG means in Practice, *AGI Conference at GIS 2001 Proceedings,* t1.10, London: AGI.

Openshaw, S. (1987) Spatial Units and Locational Referencing, reproduced as Appendix 7 in *Handling Geographic Information: Report of the Committee of Enquiry* chaired by Lord Chorley, London: HMSO.

Ordnance Survey (2001) TOID Story — How the DNF will Improve the Spatial Analysis and Web Enabling of Data, *GI News — Guide to the New Ordnance Survey,* p. 10.

Ordnance Survey (2002) *Ordnance Survey Business Portfolio 2002,* Available online at http://www.ordnancesurvey.gov.uk/businessportfolio>2002/listing.htm (accessed February 17, 2003).

Owen, R. (1999) Finding the Data Resources You Need at the Desktop: Delivery of an NGDF Metadata Service, *Mapping Awareness,* vol. 13, no. 8, pp. 46–47.

PPA (Peter Pendleton and Associates), (2003) *National Website Survey 2003,* reported in *Planning,* no. 1536, p. 2.

Raynsford, N. (1998) Minister's introduction to *Geographic Information and Public Policy: A special supplement to the Parliamentary Information Technology Committee Journal,* London: AGI.

Reeve, D.E. and Petch, J.R. (1999) *GIS Organizations and People: A Socio-Technical Approach,* London: Taylor & Francis.

Reichardt, M. (2002) New OGC Column, *GIM International,* vol. 16, no. 1, p. 17.

Rhind, D.W. and Mounsey, H.M. (1989) The Chorley Committee and handling geographic information, *Environment and Planning,* A 21, 571–85.

Rhind, D.W. (1995) Ordnance Survey to the Millenium: The new policies and plans of the National Mapping Agency, *Proceedings of AGI 95,* London: AGI.

Rix, D., Markham, R., and Howell, M. (2001) Is there a 'G' in e-Government? The Modernising Government Agenda and the Opportunities for GI, *AGI Conference at GIS 2001 Proceedings*, t2.21, London: AGI.

Robinson, A.H., Morrison, J.L., Muehrecke, P.C., Kimerling, A.J., and Guptill, S.C. (1995) *Elements of Cartography*, 6th ed., New York: John Wiley & Sons.

Roodzand, J. (2000) Spatial-IT in the Enterprise — Practical Challenges (2), *Geo Informatics*, vol. 3.2, pp. 10–13.

RTPI (1998) *1995 GIS Survey National Statistical Report*, London: Royal Town Planning Institute.

RTPI (1999) *IT Literacy Survey Summary Report*, London: Royal Town Planning Institute.

RTPI (2000) *IT in Local Planning Authorities 2000*, London: Royal Town Planning Institute.

SCST (1983) *Remote Sensing and Digital Mapping*, London: HMSO.

Serpell, D. (1979) *Report of the Ordnance Survey Review Committee*, London: HMSO.

Sommerfield, P., Corbin, C., and Mansell, J. (2000) Raising GI Awareness through Partnership, *AGI Conference at GIS 2000 Proceedings*, t2.15, London: AGI.

Sproull, L.S. and Goodman, P.S. (1990) Technology and Organizations: Integration and Opportunities, in Goodman, P.S, Sproull, L.S, and Associates (Eds.), *Technology and Organizations*, San Francisco: Jossey-Bass, pp. 254–265.

Stoter, J. (2000) Product Survey on Mainstream GIS, *GIM International*, vol. 14, no. 12, p. 49.

Thorpe, P. (1998) *Report to Enfield Council — Council Owned Property and Information Project*, Kenilworth: Peter Thorpe Consulting.

Tomlinson Associates Ltd. (1986) Review of North American Experience of Current and Potential Uses of Geographic Information Systems reproduced as Appendix 6 in *Handling Geographic Information: Report of the Committee of Enquiry* chaired by Lord Chorley, London: HMSO.

Turner, M. (2002) GI Developments in Central Government, *Geographic Information: The Newsletter of the Association for Geographic Information*, vol. 12, no. 3, London: AGI.

Tyrrell, R. (2001) National Map-maker says TOIDs R Us! *Surveying World*, vol. 9, no. 2, pp. 22–23.

UK Favourites.com. (2000) *Using New Technology to Improve Public Participation and Democracy within the Planning Service*, Online. Available <http://www.ukfavourites.com/egov.htm> (accessed September 5, 2000).

Wild, A.A. (1997) Intergraph (UK) Ltd., *Proceedings of Symposium on the Future for Geographic Information: 10 Years after Chorley*, London: AGI.

Winterkorn, E. (1994) Local Authority Usage in England, Scotland and Wales, *Association for Geographic Information 1994 Conference Proceedings*, 14.3.1–14.3.3, London: AGI.

Worboys, M.F. (1995) *GIS: A Computing Perspective*, London: Taylor & Francis.

Glossary

Accuracy the closeness of the results of observations or calculations to the true value.

Address the means of referencing a property or building.

Aerial Photograph photograph taken from an aerial platform — usually an airplane — either vertically or obliquely.

Application Service Provider a service that allows users to access data via the Web and run Web-based applications to process that data.

ARC/INFO a leading software package developed by the Environmental Systems Research Institute (ESRI).

Attribute any nonspatial characteristic of an object (e.g., total population of a zone, name of a street).

Basic Spatial Unit the smallest geographic object (spatial entity), used as a building block of a georeferencing system.

Benchmarking the process of assessing the performance of both software and hardware against a series of user-defined tests.

BS7666 spatial datasets for geographic referencing part 1 Street Gazetteer, part 2 Land and Property Gazetteer, part 3 Addresses, and part 4 Public Rights of Way.

Buffering the creation of a zone of equal width around a point, a line, or an area feature.

Business Process Reengineering radical restructuring of tasks, responsibilities, and methods as a result of introducing new technology.

Cartography the organization and communication of geographically related information in either graphic or digital form.

Cell the basic element of spatial information in the raster (grid square) format.

Computer-Aided Design (CAD) software designed to assist in the process of designing and drawing.

Coordinates pairs of numbers expressing horizontal distances along orthogonal axes, or triplets of numbers measuring horizontal and vertical distances.

Data observations made from monitoring the real world collected as facts or evidence.

Databank a set of data relating to given subject and organized in such a way that can be consulted by users.

Dataset an organized collection of data with a common theme.

Database a collection of data, usually stored in files, associated with a single general category.

Data Accuracy the extent to which an estimated data value approaches its true value.

Data Capture the process of collecting data, usually by computerized means, such as keyboard entry of text or numeric data or the digitization of spatial data.

Data Conversion the process of converting data from one format to another.

Data Editing the process of correcting errors in data input into a system such as a GIS.

Data Error the physical difference between the real world and a GIS facsimile.

Data Input the process of converting data into a format that can be used by a GIS.

Data Quality attributes of a dataset that define suitability for a particular purpose, e.g., completeness, positional accuracy, and currency.

Data Warehouse the process of assembling disparate data and then transforming them into a consistent state for decision making.

Database Management System (DBMS) computer programs for organizing information with a database at the core.

Decision Support System a computerized system dedicated to supporting decisions regarding a specific problem or set of problems.

Differential GPS the process of using two GPS receivers to obtain highly accurate position fixes.

Diffusion the process whereby technological innovations are adopted and taken up by various user groups.

Digital Elevation Model a digital model of height represented as regularly or irregularly spaced point height values.

Digital Mapping the process of storing and displaying map data in computer form.

Digital Terrain Model a digital representation of ground's surface using information on height, slope, aspect, etc.

Digitizer computer hardware used to convert analogue data into digital format.

Digitizing the process of converting data from analogue to digital format.

Entity something about which data is stored in a database, e.g., for a building the data may consist of relationships, attributes, position, or shape.

Feature Codes unique codes describing the feature to which they are attached in the GIS database.

Geographic Information System (GIS) a computer system that is designed specifically for storing, retrieving, combining, analyzing, and presenting "spatial" information.

Geographic Information Management (GIM) an integrated approach to the acquisition, definition, processing, analysis, presentation, and storage of geographic information within the organization that focuses on its practical use for business and government purposes.

GIS Champion someone who is committed to introducing GIS into an organization.

Gazetteer a geographical index.

Geocode a code that represents the spatial characteristics of an entity, e.g., the coordinates of a point or a postcode.

Global Positioning System (GPS) a system of orbiting satellites used for navigational purposes and capable of giving highly accurate geographic coordinates using handheld or backpack receivers.

Grid Reference the position of a point on a map usually expressed in terms of coordinates with the Easting distance given before the Northing.

Image a raster representation of a graphic product (scanned map, photograph, drawing, etc.) or a remotely sensed surface that consists of one or more spectral bands.

Information intelligence resulting from the assembly, analysis, or summary of data in a meaningful form.

Information System the means of storing, generating, and distributing information for the purpose of supporting operational or management functions of an organization.

Information Technology any use of equipment to provide an information system.

Internet the global network of computers originally set up as a means of secure communications for military and intelligence purposes.

Intranet the use of Internet tools to establish private "internets" within organizations.

Invitation to Tender a document forming an important step in the procurement process setting out an organization's requirements, constraints, and timetable of events together with the contractual conditions.

Land and Property Gazetteer a geographical index of land and property.

Land Terrier a set of marked maps and ledgers containing textual information to record land and property.

Layer a usable subdivision of a dataset, generally containing objects of certain classes.

Map Projection an arrangement of meridians and parallels to portray the curved surface of a sphere on a flat sheet of paper.

Mathematical Model a model that uses one or more of a range of mathematical techniques to provide a representation of reality.

Metadata information about data (e.g., dataset title, collection date, collecting organization, and data fields covered).

Multimedia a combination of a variety of user interfaces and communication elements such as still and moving pictures, sound, graphics, and text.

National Grid the metric grid on a transverse Mercator projection used by the Ordnance Survey on all postwar mapping to provide unambiguous spatial referencing in Great Britain for any place or entity whatever the map scale.

Node a point representing the start or end of a link or line or its intersection with another line.

Object-Orientated approach an approach to organizing spatial data as discrete objects for programming, modeling, and database management.

Open Systems systems that use standards to enable the operation of software from different suppliers and to share data in different formats.

Pixel the smallest element of a 2-D image to which attributes such as color and intensity can be assigned.

Point an abstraction of an object with a location specified by a set of coordinates.

Polygon an area bounded by a closed line.

Polyline a composite line formed by an interlinked set of lines.

Positional Accuracy the accuracy of the values of geographic position that can be either absolute or relative.

Postal Code (or Post Code) a code forming part of a postal address to facilitate the sorting and delivery of mail.

Primary Data data collected through first-hand observation, e.g., survey records and GPS observations.

Program a logical and connected set of instructions that tell the computer to perform a sequence of tasks.

Raster Data spatial data expressed as a matrix of cells or pixels.

Real-Time System a system able to receive continually changing data from external sources and to process that data with sufficient speed to be able to influence the sources of data, e.g., monitoring pollution levels.

Relational Database a computer database employing an ordered set of attribute values or records grouped into 2-D tables.

Remote Sensing the technique of obtaining data about the environment and Earth's surface from a distance, e.g., an aircraft or a satellite.

Resolution the size of the smallest feature that can be mapped or measured.

Satellite Image a graphic image (usually digital) taken of Earth's surface using electromagnetic sensors on board an orbiting satellite.

Scale the ratio of the distance on a map or photograph and corresponding distance on the ground.

Scale Analogue Models a model that is a scaled-down and generalized replica of reality, e.g., a topographic map or an aerial photograph.

Scanning a method of data capture whereby an image or map is converted into digital raster format by systematic line-by-line sampling.

Secondary Data data collected by another person or organization for another purpose, e.g., maps and census data.

Sieve Mapping the consecutive overlay of various maps to find areas that satisfy a given set of criteria.

Software all or part of the programs, procedures, rules, and their associated documentation of an information processing system.

Spatial relating to location or size.

Spatial Data Model a method by which geographic entities are represented in a computer; two main methods exist: raster and vector.

Spatial Reference the position of a point on the national grid expressed in terms of coordinates, the Easting (x coordinate) being given before the Northing (y coordinate)

Surface an entity type used to describe the continuous variation in space of a third dimension, e.g., terrain.

Temporal Data data that can be linked to a certain time or period of time.

Thematic Data data that relate to a specific theme or subject.

Tile a logical and rectangular set of data used to subdivide digital map data into manageable units.

Topographical Database a database holding data relating to the physical features and boundaries on Earth's surface.

Topology the spatial rules and relationships between objects including adjacency, containment, and connectivity.

Total Station a theodolite or electronic distance meter combined with a data logger and automatic mapping software.

Triangulated Irregular Network (TIN) an irregular set of height observations in a vector model connected by lines to produce a mesh of triangles that represent the terrain surface and features.

Unique Property Reference Number (UPRN) a unique identifier for a property or plot of land.

Update the process of adding to and revising existing information to take account of change.

Vector Data spatial data expressed in the form of coordinates of the ends of line segments, points, text, etc.

Virtual Reality the production of realistic looking computer-generated worlds using advanced computer graphics and simulation modeling.

Visualization the viewing of spatial data as a map or image as the basis for improved understanding.

World Wide Web (WWW) the interconnected global network of computers and the software used to access and exchange digital information and multimedia.

Index

A

ACACIA project, 50
ADDRESS-POINT, 169, 170–171, 192, 198
Address referencing, 51
Aerial photographs
  applications for, 61
  as digital databases, 49
Application data, 48
Application service providers (ASPs), 68
ArcView, 209
ArcView license, for Shepway District Council, 169, 170
Areas, measuring, 58
Association for Geographic Information (AGI), 90
  aims of, 19–20
  guidelines of, 56
  initiatives of, 91–92
  purpose of, 91
  survey of, 25
  1997 symposium of, 25
Audit Scotland, *Common Data, Common Sense* report of, 93, 94
Autodesk, 10
  as application service provider, 68
  for Shepway District Council, 169, 170
Automated Mapping and Facilities Management (AM/FM), 15
Aylesbury Vale District Council
  benefits achieved from GIS, 160, 164–165
  GIS development of, 161–162
  and GIS implementation, 160, 162, 163
  planning of, 163
  selection of, 160–161
  special characteristics of, 110, 159–166
  success of, 160, 163–164

B

Base map, 6, 7
Base scale maps, digitizing, 16–17
Basic Land and Property Units (BLPUs), coordinating significance of, 97–99
Basic spatial units (BSUs), 15
Benchmarking, 81
Boundary data, 51–52
Brainstorming, 86
Bristol City Council
  achievements of, 117–123
  benefits achieved from GIM, 126
  corporate strategy of, 121
  evaluation of GIS of, 125–126
  financial constraints on, 119
  GIS benefits achieved, 116–117
  GIS development of, 118–123
  and GIS implementation, 116
  Land and Property Working Group of, 123
  and LLC service, 120
  LPG of, 115
  NLIS pilot project of, 122, 123
  NLPG of, 122
  pioneering role of, 116, 126–127
  planning of, 124
  special characteristics of, 109, 115–116
  success of, 116, 125–126
  Web-based approach of, 124
Bristol Street Gazetteer, 118
Britain. *See also* United Kingdom
  early computer technology in, 14
  local governments in, 9
Broadband, availability of, 204
BS7666 standards, 119, 147, 178
Buffering operations, 58
Building control, in GIS applications, 133

T - #0409 - 071024 - C6 - 234/156/12 - PB - 9780367394196 - Gloss Lamination